Springer Biographies

The books published in the Springer Biographies tell of the life and work of scholars, innovators, and pioneers in all fields of learning and throughout the ages. Prominent scientists and philosophers will feature, but so too will lesser known personalities whose significant contributions deserve greater recognition and whose remarkable life stories will stir and motivate readers. Authored by historians and other academic writers, the volumes describe and analyse the main achievements of their subjects in manner accessible to nonspecialists, interweaving these with salient aspects of the protagonists' personal lives. Autobiographies and memoirs also fall into the scope of the series.

More information about this series at http://www.springer.com/series/13617

Pieter C. van der Kruit

Pioneer of Galactic Astronomy: A Biography of Jacobus C. Kapteyn

 Springer

Pieter C. van der Kruit
Emeritus Jacobus C. Kapteyn
distinguished professor of astronomy
Kapteyn Astronomical Institute
University of Groningen
Groningen, The Netherlands

ISSN 2365-0613   ISSN 2365-0621   (electronic)
Springer Biographies
ISBN 978-3-030-55425-5   ISBN 978-3-030-55423-1   (eBook)
https://doi.org/10.1007/978-3-030-55423-1

Cover picture: Jacobus Cornelius Kapteyn and his wife in 1909 in front of their tent at Mount
Wilson Observatory near Pasadena, California, not far from Los Angeles (see Fig. 9.7). From Henriette
Hertzsprung–Kapteyn's biography *J.C. Kapteyn; Zijn leven en werken* [1]. Kapteyn Astronomical Institute,
University of Groningen.

This Springer imprint is published by the registered company Springer Nature Switzerland AG
The registered company address is: Gewerbestrasse 11, 6330 Cham, Switzerland

Jacobus Cornelius Kapteyn around 1906 in a drawing by Cornelis
Easton. Kapteyn Astronomical Institute, University of Groningen.

This book is dedicated to

*The past and current staff and personnel at the
Astronomical Laboratory 'Kapteyn' and the
Kapteyn Astronomical Institute of the
University of Groningen,
and to the descendants of its founder,
Jacobus Cornelius Kapteyn.*

Preface

You know all the stars and you measure their trail
And without need at night to trace their course
You draw on paper their eternal orbits on scale
Weigh sphere after sphere: distance, mass and force.

The Universe surrounds your hermitage on this Earth
Surrounds the brain of this child of mankind,
It turns and revolves and finds no greater worth
Than that it regains itself reflected in your mind

And here in this rural home with the ones you love
Lives your heart that ensures in its solemn beat
That the passion for the aeons and suns high above
And that for wife's eye and children's laugh meet
Albert Verweij (1865–1937)[1]

Jacobus Cornelius Kapteyn (1851–1922), namesake of the Kapteyn Astronomical Institute at the University of Groningen, is regarded as the founder

---

[1]The poem is To an astronomer [2], by Albert Verweij. Verweij was a Dutch poet from the 'Movement of Eighties', translator, editor and literary historian. The original text is 'Aan een sterrenkundige//Ge kent de sterren en ge meet hun banen/En zonder hoge en nachtelijke wacht/Schouwt ge op uw blad de paden die niet tanen,/Weegt bol en bol: hun afstand, zwaarte en kracht.//Het heel heelal hangt rond uw kluis op aarde,/Hangt rond het brein van dit klein mensekind,/Het wentelt en vindt nergens groter waarde/Dan dat het in uw geest zichzelf hervindt.//En waar in het landlijk huis de lieven wonen,/Daar klopt het hart dat in zijn donkre slag/De liefde bindt van zonnen en aeonen/Aan die van vrouwenoog en kinderlach'.

of statistical astronomy and the father of the spectacular success of Dutch astronomy in the twentieth century. His *Plan of Selected Areas*, published in 1906, involved a survey in 220 small areas, distributed over the full sky, to map the distribution and motions of stars in the Milky Way. Its execution involved observatories from all over the world. Kapteyn's work was the pioneering effort that presently culminates in the European Space Agency's GAIA satellite. Kapteyn was one of the most prominent astronomers in his days, but now seems mostly forgotten. Who was this astronomer from the University of Groningen? Why is he mentioned so little in popular or academic accounts of GAIA and Galactic astronomy?

Fig. 0.1 The Kapteyn Room at the Kapteyn Astronomical Institute of the University of Groningen with Kapteyn's desk and most publications. On the right are the celestial spheres (see Fig. 11.2). On the painting by Veth (Fig. 12.3), Kapteyn sits at this same desk. On the right, the star chart, which he made as a teenager (see Fig. 2.2). It has recently been moved to another wall and replaced by the portrait of his wife as in Fig. 12.12. The books are, respectively, the *Harvard-Groningen Durchmusterung*, the *Cape Photographic Durchmusterung*, collections of his articles (Fig. 0.3), and the first volumes of the *Publications of the Astronomical Laboratory at Groningen*. Photograph by the author

One reason is unfortunate timing. Kapteyn presented his 'first attempt' at a description of the Sidereal System, as the Galaxy was usually referred to, in 1922, the year of his death. This was shortly after World War I and Kapteyn had chosen the side of those who—to the strong dislike of a majority of American and British astronomers—unsuccessfully opposed exclusion of German scientists and those from other 'defeated' countries from international scientific organizations. Furthermore, not much later, some of his work was shown to be seriously flawed due to the neglect of interstellar absorption of light, and his model for the Sidereal System was replaced by American Harlow Shapley's much larger model based on the distribution of globular clusters. This in spite of Kapteyn's earlier work that provided strong evidence for the presence (and resulted in a reasonable, although fortuitous, determination of the amount) of interstellar absorption, which he himself and other astronomers had, as we now know incorrectly, renounced a few years ago on the basis of the work by the same Harlow Shapley.

Another important reason would be the lack of a comprehensive biography of Kapteyn. It is true that in 1928 his daughter Henriette Hertzsprung–Kapteyn published a biography [1], but this was by a loving daughter, writing about her famous father with little knowledge of astronomy. A plan to write an authoritative Life by Kapteyn's student Willem de Sitter and well-known historian Johan Huizinga had come to naught. For more than 90 years since Kapteyn's death, no further account of his life and work has been written. The book by Henriette Hertzsprung–Kapteyn has been translated into English [3] by Erich Robert Paul, but the quality of this translation has been heavily criticized (Appendix B in [4]).[2]

The present book aims at restoring the appreciation of Kapteyn as the pioneer of Galactic astronomy, a field that attracts much attention these days throught the GAIA mission. It is based on a comprehensive academic version that I published in 2014 in the Springer Astrophysics and Space Science Library, *Jacobus Cornelius Kapteyn; Born investigator of the heavens* [5]. The current volume aims at a wider audience with a general interest in science. It has been published first in Dutch [6] and this is an English translation by the author.

In writing this book, I have attempted to put developments and contributions by Kapteyn in the context of contemporary astronomy. So, in Chap. 3, I have summarized the state of astronomy around 1875, the year that Kapteyn entered the field when he was appointed on the staff of the 'Sterrewacht' Leiden. At other places, where appropriate, I have summarized what was known of the subject at the time.

This volume has a sequel in a biography of Kapteyn's most famous student, Jan Hendrik Oort, again first as a scientific biography [7], and followed up by a version for a more general public, *Master of the Galactic astronomy; a biography of Jan Hendrik Oort*, which has been published at the same time as this volume in the series Springer Biographies [8].

---

[2]An improved translation by myself has been posted on my special Kapteyn Website (see below).

Sources. Of course, more has been written about Kapteyn. There are five important sources that serve as a background for this biography of Kapteyn, to which I shall refer where appropriate. In chronological order these are

- Cornelius Easton (1864–1929), journalist and amateur astronomer, Doctor *honoris causa* at the University of Groningen with Kapteyn as 'promotor', wrote an article in the Dutch amateur periodical *Hemel & Dampkring* not long after Kapteyn's death: *Personal Memories of J. C. Kapteyn*. [9]. An English translation is available in Appendix C of [5].
- As already referred to above, in 1928 Kapteyn's daughter, Henriette Mariette Augustine Albertine Hertzsprung–Kapteyn (1881–1956), published a biography of her father, *J. C. Kapteyn; His Life and Work*. My English translation is available on my Kapteyn Website (see below).
- In 1999, a symposium on Kapteyn was held on the occasion of the 385th anniversary of the University of Groningen: *The legacy of J. C. Kapteyn: Kapteyn and the development of modern astronomy*. The proceedings have been published with myself and Klaas van Berkel as editors [4].
- In 2008 the booklet *Lieve Lize: The love letters from the Groningen astronomer J.C. Kapteyn to Elise Kalshoven, 1878–1879*, by Klaas van Berkel and Annelies Noordhof-Hoorn [10], appeared. This contains letters that Kapteyn wrote to his fiancée just after he had become professor in Groningen until their marriage a year and a half later. No English translation is available. These letters were not intended for others and reading them is a bit uncomfortable; it feels like an unsolicited intrusion into the privacy of Kapteyn and his then prospective wife. However, it contains much useful information.
- My academic biography, *Jacobus Cornelius Kapteyn: Born investigator of the Heavens* [5].

To preserve their authenticity, I have used many **direct quotations** from the biography by Kapteyn's daughter Henriette Hertzsprung–Kapteyn, and from letters and other notes that I had available. Rewording or paraphrasing them would have taken away from the personal flavor and have given less insight into the personality of Kapteyn.

Website. More background to Kapteyn, such as a full list of his publications, publications about him, his academic genealogy, scans of letters from David Gill to Kapteyn, etc. can be found on a dedicated Website that I maintain: www.astro.rug/JCKapteyn. This Website also has my translation of Henriette Hertzsprung's biography.

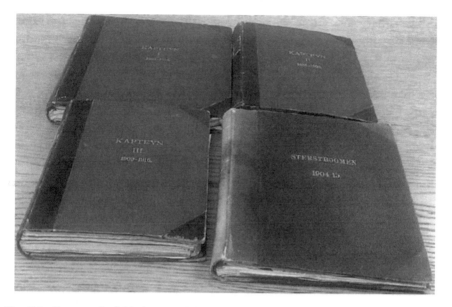

Fig. 0.2 Kapteyn had his loose publications bound as three volumes and numbered the resulting pages sequentially. In the fourth (red) volume, bottom right, he collected publications by himself and others on the Star Streams. These volumes reside in the Kapteyn Room of the Kapteyn Astronomical Institute. Photograph by the author

Persons are identified in this book by their full names and years of birth (and death when applicable) the first time they are mentioned. This then also is the first reference to that person in the Index. As in my English biography I refrain from using **nicknames**, such as 'Dody', 'Hetty' and 'Rob' for Kapteyn's children, as I feel this should be restricted to intimates.

Numbers between square brackets refer to literature sources or to Websites in the section References. Comments on these are in footnotes, but I tried to keep these to an absolute minimum not to interrupt the reading of the main text unnecessarily.

Genealogy. There are a number of useful Websites for family tree research, certificates or family messages. The ones I have used have been collected in Refs. [11]–[17].

Currency calculators. To convert currencies from other times or countries into Euro's of the present time (at least indicatively), I have used Websites, which have been collected in Refs. [18]–[20].

Geographical locations. For a map of the Netherlands in Kapteyn's days and the location of cities and other places that play a role in the book, see Fig. 1.0.

The primary aim of this book has been to document and explain the fundamental contributions of Kapteyn to astronomy in general and in the Netherlands in particular. I hope that it clarifies for non-astronomers how the field of our developing understanding of the Milky Way Galaxy in which we live for a major part goes back to Kapteyn and how it has developed at least during his lifetime. The sequel on Kapteyn's famous student Jan Hendrik Oort carries this further to the second half of the twentieth century. For readers more familiar with astronomy, I hope it provides or enhances appreciation for Kapteyn's fundamental contributions to our field. In my contribution to the Legacy' Symposium on Kapteyn that I organized in 1999 together with historian Klaas van Berkel [4], I mentioned the following. Around 1990, the Kapteyn Astronomical Institute instituted a 'preprint series'. Copies of manuscripts of papers by Groningen astronomers that had been accepted for publication by scientific journals, but which were still in press for some months, were copied and distributed to libraries of astronomical institutes worldwide. The series had a cover on which a picture of Kapteyn was displayed. Some of us on the staff were told privately and discretely by a few of our colleagues abroad (mostly in the US) that it was inappropriate to use this picture. After all, Kapteyn had been proven wrong in almost all respects. We did change this to a more 'relevant' picture related to more current research, but I now regret having not more strongly opposed that then.

In fact, Kapteyn was quite right in his earlier assessment that interstellar absorption is important and in his determination of its amount. Furthermore his analysis of the vertical structure and dynamics of the Stellar System is still essentially correct. In his *Plan of Selected Areas*, he organized an all-sky survey that made him, together with William Herschel a century earlier, one of the godfathers of GAIA. I hope that this book and the accompanying Website with links to other sources and publications will enhance the appreciation for Kapteyen as a human and as an astronomer.

Groningen, The Netherlands   Pieter C. van der Kruit
Original December 2014–October 2015,   Emeritus Jacobus C. Kapteyn
Translation November 2019–May 2020.   distinguished professor of astronomy
                                          Kapteyn Astronomical Institute
                                          University of Groningen
                                          https://www.astro.rug.nl/~vdkruit

For more background and information on Kapteyn, please consult my scientific biography [5] in the Astrophysics and Space Science Library and the dedicated Kapteyn Website www.astro.rug.nl/JCKapteyn.

Fig. 0.3 The Kapteyns lived from 1910 to 1920 at the address Ossenmarkt 6, Groningen (see Fig. 10.2). During the 1999 Legacy Symposium (*The Legacy of J. C. Kapteyn* by myself and Klaas van Berkel [4]), three of Kapteyn's great-great-grandchildren unveiled this plaque in the facade of that house. The text at the top means 'Here lived and worked 1910-1918', and under Kapteyn's name 'Inspiring Groningen astronomer'. The year 1918 (taken from his daughter Henriette Hertzsprung–Kapteyn's biography [1]) is incorrect and should read 1920.

My English translation of the French text is

'If you don't have what you love,
then you must love what you have.'

This quote is attributed to the French writer of historical articles and memoirs Roger de Rabutin, Count of Bussy (1618–1693), also known as Roger Bussy-Rabutin.

Acknowledgments

Many people have made crucial contributions to the success of this project to write a biography of Jacobus Cornelius Kapteyn. The Springer Astrophysics and Space Science Library version [5] has a more comprehensive summary and I will limit myself somewhat here.

The ASSL version was partly the result of the insistence of Adriaan Blaauw, third director of the Astronomical Laboratory in Groningen, that a comprehensive biography of Kapteyn should be written and that I should consider doing that after my retirement. I am grateful he persisted in that opinion and regret he never saw the result.

I would like to thank the 'proeflezers' (proofreaders; not in the sense of proofs of printed texts, but of the Dutch 'wijnproeven', wine tasting) for reading the original scientific biography in draft form and for commenting on it. These include my colleague and astronomer Jan Willem Pel and historians Klaas van Berkel and David Baneke. Also, my good friends Albert Jan Scheffer and Ton Schoot Uiterkamp have read and commented on the manuscript of the original book. And I also thank the proeflezers of the Dutch version for a wider audience [6]. These are my wife Corry, and my very helpful good friends Geert Hoornveld, Rob Knol and Albert Jan Scheffer. The work of all my proeflezers has been accurate and excellent; for errors that may have been overlooked, I am solely responsible.

Many thanks are also due to my editors of the previous versions, Jennifer Satten of Springer and Inge van der Bijl at Amsterdam University Press. I thank AUP also for releasing the rights on an English translation.

I am also very grateful to Springer and editor Ramon Khanna for their willingness and help to publish this biography and the accompanying one on Kapteyn's most famous student Jan Hendrik Oort. The translation from the original Dutch version has profited enormously from the availability of DeepL Translator Pro version (www.deepl.com/), which provided an excellent first step in translating the Dutch manuscript.

I would like to thank various organizations for support and help. The help from the staff of the Archives and the Museum of the University of Groningen was important and effective. This also holds for the staff of the Archives at Cambridge University (UK), the California Institute of Technology (Pasadena, California) and the Huntington Library (San Marino, California), where I had access to correspondence of Kapteyn with a number of astronomers.

For all reproductions, I have been granted permission for use in this book by the holders of the copyrights. I am grateful for their help, particulary, the Kapteyn Astronomical Institute, the University Museum Groningen and the Sterrewacht Leiden.

I am grateful to the Kapteyn Institute, my colleagues (astronomers and staff of the secretariat and of the computer group) for their help and support. I am grateful to the directors, Reynier Peletier and Scott Trager for their hospitality to generously accommodate an emeritus professor. Financial support came from the Faculty of Mathematics and Natural Sciences, in the form of an annual allowance associated with my honorary appointment to the specially created distinguished Jacobus C. Kapteyn Chair in 2003 until my formal retirement in 2009. The part of it that was left over has been used for this project. I also profited from an allowance by the directorate Exact Sciences of the Netherlands Organization for Scientific Research (EW-NWO) related to my membership of its Board. The preparation of the manuscript for this was also made possible by a further grant from EW-NWO, and from the Netherlands Research School for Astronomy (NOVA).

And finally, I thank my wife Corry for her love, understanding and support, with which she made it possible for me to write this book.

Contents

Fig. 1.0 Map of the Netherlands around 1900. The capital letters indicate the location of towns and cities with special significance to Kapteyn: B = Barneveld, where he was born and grew up; U = Utrecht, where he went to university; L = Leiden, where he first worked as an astronomer; G = Groningen, where he was a professor at the university for more than 40 years; V= Vries, where the Kapteyns owned a second house; H = Hilversum, where he and his wife bought a house to live after retirement and A = Amsterdam, where he was nursed during his fatal illness and died. Other cities and villages that occur in this book are in lowercase: al=Alkmaar, am=Amersfoort, ap=Apeldoorn, ar=Arnhem, as=Assen, bo=Bodegraven, br=Breda, de=Delft, hg=Den Haag, he=Den Helder, do=Dordrecht, ei=Eindhoven, en=Enschede, fr=Franeker, hl=Haarlem, hw=Harderwijk, go=Gouda, le=Leeuwarden, ma=Maastricht, ro=Rotterdam, vl=Vlissingen, wa=Wageningen and zw=Zwolle. Adapted from 'Kaart Nederland Jan clip art' [21]

1

Prelude and Framework

How do you start a presentation?
For example use a quote.
Taalwinkel.nl [22]

Perspective is worth 80 IQ points.
Alan Curtis Kay (b. 1940)

In the Staatscourant, the Government Gazette published by the Dutch Government announcing new laws and other governmental matters, of Saturday, December 15, 1877, a Royal Decree of the day before was published under number 36, in which it was announced that 'Dr. J. C. Kapteyn, now observator at the Sterrewacht of Leiden University, had been appointed professor in the Faculty of Mathematics and Natural Sciences at the National University of Groningen' (see Fig. 1.1). Just above that was Royal Decree number 35, in which it is announced that Kapteyn's one and a half year older brother,

---

Alan Kay, American scientist [23].

© The Editor(s) (if applicable) and The Author(s), under exclusive license
to Springer Nature Switzerland AG 2021
P. C. van der Kruit, *Pioneer of Galactic Astronomy: A Biography of Jacobus C. Kapteyn,*
Springer Biographies, https://doi.org/10.1007/978-3-030-55423-1_1

mathematician Willem Kapteyn (1849–1927), teacher at a secondary school was appointed professor at the National University of Utrecht.[1]

Jacobus Cornelius Kapteyn (1851–1922) was at that time only 26 years old, although when he took office in February 1878 he was 27 years. Figure 1.1 shows the top half of the first page of the relevant issue together with the text of the decrees, which can be found further down on the first page. Kapteyn had been born and raised in the city of Barneveld in the center of the Netherlands, where his parents ran a boarding school for boys. He had studied mathematics, physics and astronomy in Utrecht, where he had in 1975 obtained a PhD with a thesis in the area of applied mathematics. Since that he had been employed at the Sterrewacht Leiden, where he was an 'observator', which means that he was responsible for the actual carrying out of work at the telescopes. The photograph in Fig. 1.2 shows Kapteyn and his parents, brothers and sisters, and their partners. Kapteyn and his brother Willem appear just to the right of the middle with another brother wearing a hat in between them; Jacobus is the one on the left of these two; his later wife is not in the picture yet.

This appointment of Jacobus Kapteyn was a direct consequence of the Higher Education Act, which had been passed the year before. This law changed the primary task of the university: from an educational institute where young men were educated to be scholars with knowledge of science, it became an institution where research was carried out alongside scientific teaching. As a result, each of the three government-funded universities would have its own professor of astronomy. The number of universities in the Netherlands was also increased to four; while a few years earlier it was expected that one of the three state-financed universities at Leiden, Groningen and Utrecht would be closed—and the expectation was that it would be that of Groningen. Not only did all three continue, but the Amsterdam 'Athenaeum Illustre' was finally granted the status of (municipally financed) university. Amsterdam did not get a professor of astronomy and it was not until 1921 that an astronomy lecturer was appointed. The number of subjects taught in academic education, government funding of universities and the number of professors all roughly doubled.

---

[1]The universities of Leiden, Groningen and Utrecht were a 'Rijks-universiteit' (funded by the national government), sometimes translated as 'State University' (although 'National University' would be more appropriate). Leiden and later Utrecht in the 1990s dropped the 'Rijks' prefix, because they felt the translation 'State University' would suggest mediocrity. In Groningen, the Rector Magnificus and Deans of the Faculties—of which I was one—felt the 'Rijks' was appropriate and chose to keep it, but decided that 'University of Groningen' would be the official English name.

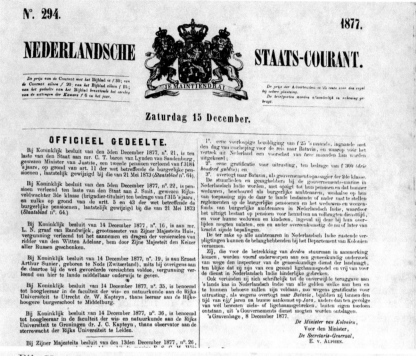

Fig. 1.1 The heading and upper part of the front page of the *Staatscourant* (the Government Gazette) of 15 December 1877 (top). Below the text of the decisions to appoint the Kapteyn brothers as professors

Although astronomy had been taught in Groningen since the founding of the university in 1614, there had not been a special chair in it until then. This was in contrast to Leiden and Utrecht, where respectively Hendricus Gerardus van de Sande Bakhuyzen (1838–1923) and Jean Abraham Chrétien Oudemans (1827–1906) not only taught astronomy—but also mathematics and physics—, but also had their own observatory for scientific research. In Groningen there were no facilities for astronomical education, let alone research. Kapteyn would never see his fervent desire to have his own observatory realized. This is the background to the quote, which was close to his heart: 'If

Fig. 1.2 Reunion of the Kapteyn family. The date is not specified. We see that father G.J. Kapteyn, in the front (fourth from the left), is still alive and that Catharina Elisabeth (Elise) Kalshoven, Kapteyn's future wife, is not in the picture. This puts it at around 1878. Jacobus Kapteyn is in the back row just to the right of the center. (Fotocollectie gemeentearchief Barneveld) [24])

you don't have what you love, you must love what you have', which is now placed in the facade of the house on the Ossenmarkt in Groningen where he lived for about ten years (see Fig. 0.4).

The practice of astronomy in the Netherlands[2] dates from the seventeenth century, not long after the establishment of the first university in the Netherlands. Leiden University was founded in 1575. This was even before the northern Seven Provinces of the Low Countries (or Spanish Netherlands) had declared themselves no longer subordinate to the King of Spain with the 'Plakkaat van Verlatinghe', the Act of Abjuration, of 1581, and the founding act of Leiden University consequently still carries the seal of Philips II. It is ironic that the University of Leiden, which carries the motto 'Praesidium Libertatis', bastion of freedom, was legally founded under the rule of the King of Spain. In 1633 the University was given an observatory, and it claims to have the oldest university observatory in the world. It was initially housed on the roof of the Academy building.

---

[2]A more extensive, very readable description can be found in the introductory chapter of David Baneke's recent book on Dutch astronomy in the twentieth century [25].

The University of Utrecht, founded in 1636, followed by founding an observatory in 1642, not much later than that in Leiden. It was located on the 'Smeetoren' (Smith Tower) on the ramparts, which had been used by the smiths' guild. Unlike in Kapteyn's days, these observatories were originally mainly used for education and demonstration (to special guests). No observatory was set up at Groningen University, but from the very beginning there were lectures in astronomy, starting with Nicolaus Mulerius (1564–1630).

When the Netherlands became a Kingdom in 1815, the observatories of Leiden and Utrecht were far from 'state of the art'. Teaching in astronomy was often provided for by professors for whom this was not their main task nor their primary interest. That changed in the middle of the nineteenth century. In Utrecht, the catalyst was Christophorus Henricus Didericus Buys Ballot (1818–1890), best known as a meteorologist, who not only lectured in physics but also in astronomy. In 1854 he succeeded in establishing a meteorological institute on the 'Sonnenborgh', on the city walls; he had already arranged for the astronomical observatory to move there. As early as 1843, Adolf Stephanus Rueb (1806–1854) had been appointed a lecturer to release Buys Ballot from his many tasks and to provide astronomical education. After his death in 1854, Jean Oudemans was attracted as extraordinary professor (additional to the structural or ordinary professors). Oudemans left in 1857, but returned in 1875 as professor of astronomy and director of the Utrecht Observatory.

In Leiden, people did not sit still either. In 1826, Frederick Kaiser (1809–1878) was appointed as observator—a person carrying out the observing. He made quite a name for himself by predicting the return of Halley's comet in

Fig. 1.3 The Sterrewacht Leiden in 1865, a few years after it had been completed. The founder Frederick Kaiser had modeled it after the Pulkovo Observatorium near St. Petersburg. Kapteyn worked at the Sterrewacht between 1875 and 1877 as observator. Archives Sterrewacht Leiden

1835 more accurately than many other astronomers, and was given a new telescope and an observator as assistant (the same Oudemans who later went to Utrecht). He then became a professor in 1840. In 1861 a brand-new 'Sterrewacht' was opened in the hortus botanicus. Figure 1.3 show a picture of the Sterrewacht Leiden in 1865, not long before Kapteyn worked there. Kaiser retired in 1872 and Hendricus van de Sande Bakhuyzen succeeded him as director and had been responsible for hiring Kapteyn as observator in 1875. Kapteyn worked here until he was appointed in Groningen early 1878.

At the time of Kapteyn's appointment in Groningen, Dutch astronomy played a modest role internationally. The Sterrewacht Leiden had acquired an excellent reputation thanks to the work of Frederik Kaiser and Hendricus van de Sande Bakhuyzen because of the great accuracy with which positions of stars and other objects were measured. But other countries such as Germany and the United Kingdom set the tone. Despite, and perhaps thanks to, the handicap of not having his own observational facility, Kapteyn became in his days one of the most influential and leading astronomers in the world. Since then, Dutch astronomy has occupied a prominent position worldwide. The Dutch Research School for Astronomy (NOVA), a coordinating organization of the research institutes at Dutch universities, was selected in 1998 as one of six 'top research schools' in the Netherlands. In an international assessment in 2010, two of the top research schools subsequently received the designation 'exemplary' (even higher than 'excellent'). NOVA was one of the two (and this status has been confirmed since) and astronomy is therefore among the absolute top of Dutch scientific endeavor.

That enormous success started with Kapteyn. To understand the way this came about we need to go back to the middle of the nineteenth century to the small town of Barneveld, where Jacobus Cornelius Kapteyn was born in 1851.

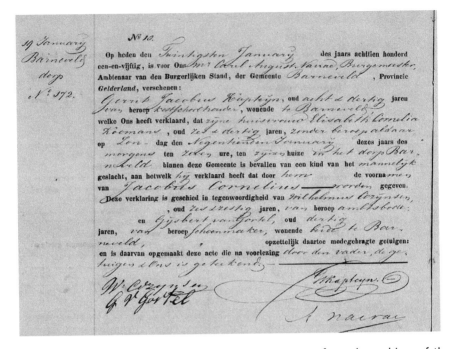

Fig. 2.0 The birth certificate of Jacobus Cornelius Kapteyn from the archives of the municipality of Barneveld. The birth, which took place at the parental home, is dated Sunday January 19, 1851, at 7 o'clock in the morning. Witnesses (to the registration of course) were Wilhelmus Corijnsen, aged 66, a municipal messenger—'Ambtsbode'; a person employed by the municipal administration, whose primary task was to deliver messages and documents—, and Gijsbert van Gortel, aged 30, a cleaner by profession. Note that here the last name is spelled Kapteijn. From FamilySearch [26]

2

Boarding School and University

You will never finish with your family,
that is just like chicken pox, which you get in your youth
and throughout your life it leaves its mark.
Jean-Paul Sartre (1905–1980)

Mathematics is the gateway and the key to all science.
Roger Bacon (c.1214–1294)

The middle of the nineteenth century was a time of reform in the Netherlands. In 1848, King Willem II (1792–1849), who had succeeded his father in 1840, feared that the various popular uprisings in Europe would spread to the Netherlands. To turn the tide, or at least to be ahead of it, he invited the Second Chamber (House of Representatives) of Parliament to set up a committee to prepare a major revision of the Constitution. This committee was headed by Johan Rudolph Thorbecke (1798–1872), who was a follower of liberalism, and gave its advice so quickly in the form of a draft revision, that it already came into force later that same year as the Constitution of 1848. The power came largely in the hands of the people's representatives. Ministers became responsible for policy and administration and also for the actions of the King, who became inviolable. From now on, the Parliament was elected by the people,[1] the House of Representatives directly and the Senate by the Provincial Councils. The Netherlands became a Constitutional Monarchy, and still is.

---

[1]That is to say only men voted, and then not even all of them. Somewhat later, persons graduating from a university automatically obtained voting rights. It took until 1917 before all men would have the right to vote; women followed in 1919.

© The Editor(s) (if applicable) and The Author(s), under exclusive license
to Springer Nature Switzerland AG 2021
P. C. van der Kruit, *Pioneer of Galactic Astronomy: A Biography of Jacobus C. Kapteyn*,
Springer Biographies, https://doi.org/10.1007/978-3-030-55423-1_2

9

The liberal movement, led by Thorbecke, advocated a society with as much freedom as possible for individuals. The rule of the state and of the church had to be separated and limited to what was necessary. Freedom of religion was enshrined in the constitution and turned the effectively Protestant Netherlands into a secular state. In 1853, Thorbecke allowed the Roman Catholic Church to re-establish dioceses and appoint bishops. The change was particularly great in the field of education. Until then, public schools had in fact provided Protestant education, which led to discontent among Catholics and Jews. For the establishment of schools next to the public ones, permission had to be requested and when approved in general no subsidy from the government was made available. From now on everyone was free to found a school according to their own religious beliefs. This was an important but not yet decisive step in the 'schoolwar', because the special schools were disadvantaged in a financial sense. The Education Act of 1857 regulated a system of primary and comprehensive secondary education and the establishment of special schools in addition to the public schools, an option which Catholics and Jews, but also orthodox protestants, began to use extensively. Traditional protestants were not all that happy with these developments.

Thorbecke led three cabinets, in the beginning characterized by opposition from King Willem III (1817–1890) (who, more than his father, had problems with his new constitutional role) and the moderate Protestants. In 1863, during his second term as prime minister, he reorganized secondary education. A crucial developments was that he set up a new type of secondary school in addition to the gymnasium. The latter was the route to prepare for university studies. This new type of school was called the 'Hoogere Burgerschool' (literary Higher Civil School) HBS, which was set up with the aim of educating boys (not yet girls) from the bourgeoisie, preparing them for the more influential positions in society, particularly in trade and industry. This would ultimately have a major impact on the academic world as well. In the first place because at the HBS teachers had an academic background, often with a PhD degree. This was one major way of offering a career with standing to academics with a science background. Because of this, the HBS grew into a solid secondary education in particular in mathematics, physics and chemistry and the modern languages English, French and German. The establishment of the HBS was an important stimulus for our 'second Golden Age' and is at the root of the success of Dutch science at the beginning of the twentieth century and the relatively large number of Nobel prizes won by Dutch scientists in the early decades. The Nobel Prize winners in physics and chemistry Jacobus Henricus van 't Hoff (1852–1911) in 1901, Hendrik Antoon Lorentz (1853–1928) and Pieter Zeeman (1865–1943) in 1902, Johannes Diderik van der Waals (1837–

1923) in 1910 and Heike Kamerlingh Onnes (1853–1926) in 1913, all except Johannes van der Waals, came from the HBS.

As a final step, but after Thorbecke's death, higher education was also reformed with the above-mentioned law of 1876, in which the number of universities was increased to four and the curricula offered a wider range of subjects and disciplines.

The Netherlands was divided geographically by religious conviction. At the beginning of the nineteenth century, Catholics dominated in the south-eastern half of the country, and in some areas in the present 'Randstad', the area of the four major cities Amsterdam, Den Haag, Rotterdam and Utrecht [27]. The border area between the predominantly reformed (the north-western part) and the predominantly catholic part eventually became the Bible Belt or Refoband, which was dominated by an strict and orthodox protestant population. It is unclear, however, whether there is a causal connection and to what extent it already existed in the middle of the nineteenth century. A place like Barneveld, which lies in the Refoband, was predominantly reformed or otherwise protestant, but there was not (yet) a sharp separation between orthodox and liberal. The current Refoband is characterized by a relatively large percentage of orthodox, 'bevindelijk'[2] reformed protestants, but the Refoband itself could probably not yet be recognized.

2.1   Barneveld

The city of Barneveld lies very close to the geometrical center if the Netherlands. By the middle of the nineteenth century, Barneveld had about three thousand inhabitants. The village was then predominantly reformed protestant, but there was not (yet) a sharp separation between orthodox and liberals. It was a center of egg production; around 1850, the 'Barnevelder' was bred as a cross between laying hens and large Asiatic breeds, giving rise to birds with an excellent production, laying of up to 180 eggs per year. Barneveld is known from the history books as the place where Jan van Schaffelaar jumped off the tower in 1482. This was during the 'Hook and Cod Wars', which comprise a number of battles between 1350 and 1490 over who was entitled to be the 'Count of Holland'. These 'Hoekse en Kabeljauwse Twisten' were fought between two fractions, one calling themselves the 'Cods', presumably a re-appropriation related to the arms of Bavaria, that supported Margaret, wife of the emperor of Bavaria, and the 'Hooks'. The latter is probably related to the

---

[2]Bevindelijk comes from 'bevinden', experience. The followers emphasize the importance of a personal experience of salvation by Jesus Christ.

hooked instrument used in catching cod. When Philips the Good of Burgundy appointed his son as bishop of Utrecht, this was opposed by the Hooks. The Cod van Schaffelaar and his men hid in the tower of Barneveld, which was subsequently besieged by the Hooks. Rather than surrender, he bought free passage for his men by jumping from the tower.

Barneveld was also known as a place with a number of excellent boarding schools. Prominent among these was the boarding school for boys, *Benno*, which was run by the husband and wife Kapteyn–Koomans. The name Benno is derived from German and has the original meaning 'strong as a bear'. Jacobus Cornelius Kapteyn was born there on January 19, 1851.

We are in the fortunate circumstance that Kapteyn's second daughter, Henriette Mariette Augustine Albertine Hertzsprung–Kapteyn (1881–1956), wrote a biography of her father ([1], see also the Preface). This book contains information about Kapteyn that he himself will have told her, but also memories of his wife (and Henriette's mother) Catharina Elisabeth (Elise) Kapteyn–Kalshoven. Henriette wrote the biography in 1928; she then lived separated from her husband, the astronomer Ejnar Hertzsprung (1873–1963), in Hilversum, in the same street as her mother; it is obvious that they must have regularly discussed the progress of the biography, while Henriette was working on it.

Gerrit Jacobus Kapteyn (1812–1879) and Elisabeth Cornelia Koomans (1814–1896) were married in 1837 in their birthplace Bodegraven. Gerrit Kapteyn was a teacher and came from a long family tradition of educators. He had obtained all the formal qualifications that existed in his profession, which was quite unusual. The couple started a boarding school in Voorschoten, near Den Haag, but after a few years they had to give up because they could not keep the 'Haagse' boys in check, who, accustomed to luxury, were difficult to control and educate. Gerrit Kapteyn successfully applied for a job at a school in Barneveld. After a few years he and his wife founded their own boarding school for boys (see Fig. 2.1), which quickly became famous. In Voorschoten they had already had two sons and in Barneveld another thirteen children were born. Jacobus Cornelius was the tenth in the series of fifteen. He was named after his aunt Jacoba Cornelia Kapteyn, who was called 'Aunt Ko'; to his annoyance Jacobus was also called 'Ko', a name he apparently hated greatly. His daughter describes him in her biography [1] as follows:

Jacobus Cornelius was a delicate child and the only one that gave his parents reasons for concern for his health, although he never was seriously ill. He was slender built and pale and had a high voice. In bearing and complexion he made a absent-minded and introverted impression. His long upper eyelids reinforced that impression.

Fig. 2.1 The boarding school *Benno* in Barneveld. This illustration comes from Henriette Hertzsprung–Kapteyn's biography [1]

The Kapteyn children received the same treatment as the boys who attended the boarding school; Jacobus later emphatically complained about the great lack of love and attention from his parents, who were too busy caring for the large group of boys. The Kapteyn children did not give each other any special attention either; they treated each other like the other pupils, so that there was no special bond between the brothers and sisters. The atmosphere at the boarding school was stern and religious; there was a strict regime and people studied hard and seriously. The old Kapteyn set himself the goal of shaping his students 'with the blessing of the Lord into virtuous and able men'. Every day started at seven o'clock with a reading from the Bible before breakfast.

Jacobus Kapteyn proved to be an excellent pupil. He went through primary school (at the school of his parents) in a short period of time and turned out to have special talents. In *J.C. Kapteyn; His life and works*, his daughter tells us that, as a boy of about ten years old, he was watching a game of chess between his father and elder brother Hubert. The latter invited the boy to play a game as well: Jacobus defeated his older brother in a quick game. And then a second and a third time. The brother from then on refused to play chess with him. And when a friend of the same brother once offered Jacobus to help with a mathematical exercise, he turned out to be well ahead of this friend. Such incidents happened more often and helped him to feel more and more isolated and lonely. It undoubtedly explains, at least in part, the lack of friendship and attention that he experienced during his childhood. His shyness also prevented him from feeling at ease in the company of girls.

Kapteyn spent a lot of time in the library of the boarding school and developed, among other things, a great interest in animals. Throughout his life he would entertain an extensive knowledge of and attention for birds. He recognized birds not only by their appearance, but also by their call or song. He also kept rabbits; he earned money for their food by selling pretzels at school and picking up the newspaper for the minister and delivering it to him. Henriette Hertzsprung–Kapteyn tells in her biography [1] that he once caught an owl and kept it in the attic. He had heard that owls were blind by day and wanted to check that out. He stretched wires through the room, which the owl, when released, could easily evade. Kapteyn might have become a biologist or ornithologist if he had not, at the age of fourteen, been given a star chart or globe by his older sister Albertina Maria (Bertamie) (1847–1927), who had been to England. With that, one could identify constellations at night. The gift inspired him to create his own star chart, which he did with great dedication and effort. He carefully cut the brightest stars (of the first magnitude)[3] out of gold-colored paper, the ones of the second magnitude out of silver paper and the one of the third he drew accurately with white paint. This map (see Fig. 2.2) is now located in the Kapteyn Room in the Kapteyn Astronomical Institute of the University of Groningen. Kapteyn's interest in the starry sky was aroused and he read in his father's library anything about the stars he could lay his hands on. In Henriette Hertzsprung–Kapteyn's biography [1] we read that his father noticed this interest and bought him 'a big telescope' and set it up in the tower room. This is probably the structure at the back of the picture in Fig. 2.1. A brochure from that time in the archives of the municipality of Barneveld also mentions the presence of an 'observatory' at the boarding school.

As a young teenager, Kapteyn thought a lot about social reforms. His daughter says in her biography that he talked about this with his mother, who was, however, far too practical for that. She would have said, 'And now would you like to help reform the world?'. To which young Kapteyn responded: 'Yes, Mother, if I can with my limited powers get out of the way just one little straw, I will be satisfied.' At the same time, he must have come to doubt the faith of his parents. He could not accept the dogmatism of it, the lack of inspiration and warmth. For him, faith in his own freedom and respect for the freedom of others assumed a central position. He acquainted himself with other religions; for example, he knew a Jewish boy, with whom he took part in the Feast of Tabernacles. He developed a vision in which people (at least, men) of all ranks and positions should be treated in the same way.

In the biography of the Groningen ecclesiastical professor of practical theology, Aart Jan Theodorus Jonker (1851–1928) by Maarten van Rhijn (1888-1966) [28], we come across a passage about Kapteyn. Jonker had retired in 1909

---

[3] See Sect. 3.2 for the definition of magnitudes of stars.

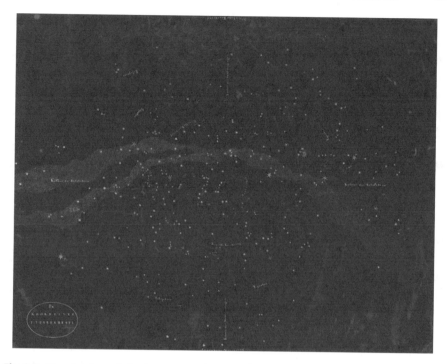

Fig. 2.2 Star chart, made by Kapteyn at the age of fourteen; it now hangs in the Kapteyn Room at the Kapteyn Astronomical Institute in Groningen. Kapteyn Astronomical Institute, University of Groningen

and had then settled in Heerde, but old and younger friends from Groningen still visited him there regularly. Maarten van Rhijn wrote:

> When one of his younger friends once told him that the astronomer Kapteyn liked to read the Saint John Gospel in his spare time, Jonker asked him what Kapteyn actually believed himself. In response to the answer: 'Kapteyn believes that everything will probably be fine in the end', Jonker immediately replied with the words: 'Jeepers, that's an excellent faith'.

Maarten van Rhijn, himself a professor of theology in Utrecht, was a brother of Pieter Johannes van Rhijn (1886–1960), eventually Kapteyn's successor as professor of astronomy in Groningen. It is quite possible that this 'younger friend' actually was Pieter van Rhijn (or Maarten himself, who had heard it from Pieter). I thank Klaas van Berkel for bringing this to my attention.

Kapteyn was only sixteen years old when he passed the entrance exam for the University of Utrecht. However, his father thought he was far too young to go and live in the student town and he stayed in Barneveld for another year. He spent his time reading books from his father's library, such as Goethe, Rousseau and others. Throughout his life the play *Nathan the Wise* by Gotthold Ephraim Lessing (1729–1781), a plea for religious tolerance that was forbidden by the church, remained one of his favorites. He also read the work of the poet and Anabaptist Robert Hamerling (1830–1889) and of the leader of the Munster Anabaptists,[4] Bernhard Knipperdolling (1495–1536). But he also read textbooks, so that when he actually went to Utrecht a year later, he had already mastered a large part of the material for first-year students. He used the math book that his older brothers also had used, *Cours d'analyse de l'École Polytechnique* by Charles-François Sturm (1803–1855). He actually developed his own way of solving mathematical problems, which later caused him problems with his mathematics professor in Utrecht, Grinwis, who insisted on using the standard methods.

Kapteyn's decision to distance himself from the faith and the church later brought him into great conflict with his parents, especially his father, when he decided not to marry in the church (I will go into more detail on this in Sect. 4.4). His parents refused to give their permission, which raises the question of how orthodox in their faith they actually were. It is unlikely that they belonged to the most orthodox movement within the church. Kapteyn Sr. is listed in Barneveld Archives as belonging to the Reformed Church. But how orthodox a church this was at that time depended to a large extent on the minister who preached there. The question of whether a church was just reformed or orthodox will provide less of a definite answer than one would now suspect. If father Kapteyn had been extremely orthodox, many parents would not have sent their sons to his boarding school, because they would then not have been educated there in the atmosphere of the State Church. Moreover, an orthodox man would not easily include the books of liberal writers such as Rousseau or Lessing in his library. Nevertheless, Kapteyn's parents were certainly very strict in their teaching and had great difficulty with the fact that their son turned his back on the church.

2.2 Utrecht

The Netherlands was relatively late in setting up universities. The southern Netherlands already had a university in Leuven in 1425, but in the northern provinces that of Leiden was, as mentioned above, the first in 1575. Not long afterwards more universities followed, Franeker in 1585, Groningen in 1614, Utrecht in 1632 and Harderwijk in 1648. In Amsterdam there had been an

---

[4]Anabaptists oppose the taking of oaths and ecclesiastical practices such as the baptism of children.

Athenaeum Illustre since 1632, but this institution was not recognized as a university and no PhD degree could be awarded there. Franeker's university no longer exists, but it flourished until the eighteenth century, although by the end of this period the number of students was low. The University of Harderwijk had a reputation as a rather cheap university to obtain a degree, which physician and botanist Herman Boerhaave (1668–1738) had done, as well as the Swede Carl Nilsson Linnæus (1707–1778), designer of the systematic classification scheme for the plant and animal kingdoms. The latter appears not to have been in Harderwijk for more than two weeks, just enough to have his thesis printed. During the French occupation, all the universities degraded to *école secondaire* or similar status, except for those of Leiden and Groningen. Only after the founding of the Kingdom of the Netherlands, by the so-called 'Organiek Besluit' (Organic Decree) of 1815, the University of Utrecht was re-established. Those of Franeker and Harderwijk were continued as atheneums, but turned out not to be viable and were closed in 1818 and 1843 respectively. As mentioned above, it was not until 1876 that Amsterdam was awarded a (municipal) university. According to the 'Organic Decree', the universities in Groningen, Leiden and Utrecht each had five faculties: theology, law, mathematics and physics, medical sciences, and philosophy and literature. Within mathematics and physics four professors taught mathematics, physics, astronomy, chemistry, biology and agriculture.

Kapteyn received his education from the professors Buys Ballot and Cornelis Hubertus Carolus Grinwis (1831–1899) (see Fig. 2.3). Grinwis taught mathematics, Buys Ballot physics and astronomy and of course meteorology, of which he is the founder in the Netherlands. Together with the physiologist and authority in the field of eye diseases Franciscus Cornelis Donders (1818–1885), biologist and early evolution and Darwin supporter Pieter Harting (1812–1885) and chemist and protein expert Gerardus Johannes Mulder (1803–1880) they probably formed the most renowned faculty of exact sciences of their time in the Netherlands. Utrecht was an excellent choice for those who wanted to study mathematics and physics, but Kapteyn's choice to go there was most likely also a result of the proximity of Barneveld and especially the fact that his older brothers also studied in Utrecht.

Two of those brothers had preceded him with a study in mathematics and physics in Utrecht and both would, like Kapteyn, obtain a PhD with Grinwis as supervisor. These were Nicholas Pieter (1845–1916) and Willem. We have already met the latter as the one who was appointed professor of mathematics in Utrecht at the same time as Jacobus was appointed in Groningen. Not only did they defend their theses with the same professor, they also did so on the same day, 14 June 1872. Willem's dissertation was an example of applied math-

Fig. 2.3 Christophorus Henricus Didericus Buys Ballot and Cornelis Hubertus Carolus Grinwis. University Museum Utrecht, with permission

ematics: *On the theory of vibrating plates and its connection with experiments*; that of the older brother concerned a purely mathematical subject: *On the calculus with symbols and their application to the integration of differential equations*. Nicolas Pieter received the judicium (qualification) *magna cum laude*, Willem *cum laude*. The three brothers, who partly studied simultaneously in Utrecht (Jacobus started his studies there in 1868), seem to have had little contact with each other, much in the tradition of family life at the boarding school in Barneveld. This despite the fact that the brothers Willem and Jacobus also lived at the same address in Utrecht at the Predikherenkerkhof [10].

Kapteyn's studies went well, although Henriette Hertzsprung–Kapteyn in her biography [1] tells us that at first he did not work too hard and surrendered to student life. Unlike his brother Willem, Jacobus became a member of the Utrecht student fraternity. An important friendship he made there was with Henry William Andrée Wiltens[5] (1851–1917). It was in this friendship that Kapteyn experienced real affection and warmth for the first time in his life, and that always stayed with him. Kapteyn's first exam did not go smoothly. He was asked about a subject of which he knew little. Such an exam, probably the 'kandidaats' exam or something similar (so now the bachelor degree), was public and everyone, including father Kapteyn, was very worried about the outcome. Jacobus, however, suggested to the professor that it would be better for them to talk about something else, since he did know nothing about this

---

[5]Andrée was part of the surname. Kapteyn referred to him as Willy Wiltens.

Fig. 2.4 Jean Abraham Chrétien Oudemans and Martinus Hoek (1834–1873). University Museum Utrecht, with permission

part. The professor did and Kapteyn in the end passed with flying colors. In *J.C. Kapteyn; His life and work* his daughter told about the party afterwards in the evening, when the old Kapteyn fell out of his role (there was good reason he was nicknamed 'Dux', Latin for captain/leader, the source of the English 'Duke') and told stories with his legs on the windowsill.

What about astronomy in Utrecht? As early as 1642, a few years after the founding of the university, an observatory was set up on the 'Smeetoren' (Smith Tower), which had previously been used by the smiths' guild. Astronomy was been taught in Utrecht since then. In 1843, Adolf Rueb was appointed lecturer in astronomy. When he died, the teaching of 'star science' was taken over by Buys Ballot, then a professor of mathematics, later of physics. In 1854 he had established the meteorological institute on the Sonnenborgh, where under Rueb the astronomical observatory was also set up. In 1856 another extraordinary[6] professor of astronomy was appointed, Jean Oudemans (Fig. 2.4, left), but, as already told above, in 1857 he accepted a position as a surveyor in the Dutch East-Indies, and in 1859, after Professor Oudemans' departure, Martinus Hoek (1834–1873) (Fig. 2.4, right) was appointed professor of astronomy. Hoek occupied this post when Kapteyn came to study in Utrecht. Hoek did not have much interest in practical astronomy. He occupied himself with physics experiments. Among other things, he did an early version of the well-known Michelson–Morley experiment, in which, by means of measure-

---

[6]Extraordinary professors were appointed in addition to the structural professors in order to provide expertise in a particular subject that was missing or to relieve a professor from an excess of duties.

ments of the propagation of light in relation to the motion of the Earth, an attempt is made to show that ether is the medium in which light propagates. However, Hoek died in 1873, at the young age of 39.

Because of Hoek's lack of interest in practical astronomy, Kapteyn could not have learned much from him. Moreover, he was no longer alive at the time when Kapteyn was looking for a subject for a PhD thesis. After Hoek's death, Buys Ballot had temporarily taken on responsibility for astronomical education again, but at the time was not interested in supervising a PhD in astronomy. He has actually supervised only one PhD thesis on an astronomical subject and that was in 1871 on the determination of the orbit of an asteroid. Jean Oudemans returned to Utrecht as a professor of astronomy, but that was in 1875, too late for Kapteyn, who already completed his dissertation in that year under Grinwis.

2.3 Vibrating Flat Membranes

Kapteyn's doctoral research was a sequel to that of his brother Willem (see Fig. 2.5). The latter had, also under the supervision of Grinwis, researched the theory of vibrating plates. This is the phenomenon, first studied exhaustively by the German physicist Ernst Florens Friedrich Chladni (1756–1827), that if you sprinkle a powder on a flat metal plate and then make it vibrate, for example with the aid of a bow for a violin or so, you get to see a regular pattern. This pattern shows where 'nodes' and 'antinodes' are in the vibrations, which indicate where the vibration is greatest (the antinodes) and where absent (the nodes). Compare this with a vibrating string: in the middle the vibration is maximal and the ends of the string are stationary. In the first overtone the tone is twice as high and the string vibrates like two half strings attached to each other. Then halfway there is no vibration (a node) and at one and three quarters the amplitude it is maximum. The theory of how that pattern is determined by the shape and elastic properties of the plate had already been studied by great scholars, such as Jacob (II) Bernoulli (1759–1789), Joseph-Louis Lagrange (1736–1813), Simeon Denis Poisson (1781–1840), Augustin Louis Cauchy (1789–1857) and Gustav Robert Kirchhoff (1824-1887), but also by the much less well-known French female mathematician Marie-Sophie Germain (1776–1831).

Willem had not done much more than discuss the work of earlier researchers; he concluded that Germain had started with a wrong assumption, but had nevertheless derived the correct equation. He also discussed the experiments

Fig. 2.5 The title pages of the dissertations of Willem and Jacobus Cornelius Kapteyn

and concluded that theory and experiment were in agreement. He did not add much new. Maybe that is why the judicium was 'only' *cum laude*.

Jacobus Kapteyn did his follow-up research with the special case of membranes instead of plates (Fig. 2.5). The thickness of membranes can be left out of consideration (Willem had already limited himself to the planes halfway inside the plates) and that made the research easier. One can also stretch membranes between a frame, while plates need to be supported. Many results of experimental work were already available for square and rectangular, triangular and circular and elliptical membranes. Kapteyn's derivations for the shape and positioning of the positions of nodes are described in a very clear manner, although it is also clear that it is mainly a presentation of other people's earlier work. But there is probably more: what seems new here is that Kapteyn predicted that there are not only lines of nodes, but that point-like nodes must also occur. In the experiments these are not visible, which he reasonably attributes to irregularities in the thickness and elasticity of the membranes. I present the research in a little more detail in Appendix A.1.

The research is interesting, but what is particularly striking are the propositions added to Kapteyn's thesis. It is and was a feature of Dutch PhD theses, that they are accompanied by a set of propositions that have to be defended too at the ceremony. In total there were eighteen, of which the numbers 10 to 16 are astronomical in nature—a clear indication of his strong interest in this field. Among the non-astronomical ones there was a remarkable proposition, because of the person involved, Dutch author Multatuli (pseudonym for Eduard Douwes Dekker, 1820–1887). The proposition deals with the use of the 'zero' in roulette. The details are not so important here. But the fact that he quoted Multatuli shows that he had read his work. It is known from other sources that Multatuli was an important source of inspiration for Kapteyn and that Multatuli's non-conformist view of society and religion appealed to him particularly. In Henriette Hertzsprung–Kapteyn's biography [1] Multatuli's work does not appear in the inventory of the library in Barneveld, so Kapteyn may have read Multatuli later during his student days. Nevertheless, Multatuli will have influenced Kapteyn's development and almost certainly played a role in his rejection of his parents' religion.

I will go into two of the astronomical propositions in more detail. Number X reads: *The best photometer is that of Zöllner.* A photometer is an instrument to measure the brightness of a star in the sky. The proposition suggests that Kapteyn was aware of the properties of such photometers and had experience using them. Johann Karl Friedrich Zöllner (1834–1882) had developed a stellar photometer in 1858 (see Fig. 2.6), which compared a star with an image of a flame from a so-called Bunsen burner [30, 31]. This is designed to give a gas flame that is particularly stable and not flickering. One could then adjust the brightness of the resulting image with a polarizer to match that of the star. Light can be interpreted as a wave and can be polarized when it passes through a transmitting thin plate, which, due to the orientation of the crystals, only allows light to pass that oscillates in a certain plane. A second plate is then placed behind it; if it is 'parallel', it allows all the light to pass through, but it can be rotated, so that only a part or nothing passes through. In this way one can adjust the brightness of the 'artificial star' to that of the star of which the brightness has to be determined. Of course, one must first observe a star of known brightness to calibrate the instrument. In Sect. 10.3 a photograph of Willem de Sitter has been reproduced (Fig. 10.11), where he is depicted next to a telescope with a Zöllner photometer mounted on it.

Until then, the brightness of stars was measured by means of visual comparison with one or more other stars. John Frederick William Herschel (1792–1871) used what he called an 'astrometer', with which he compared a star with a downsized image of the Moon in his field of view in the telescope, which he obtained after reflection in a prism and focusing with a lens. Another method was that of Carl August von Steinheil (1801–1870) in München, who had developed a device with which he compared the brightness of out-of-focus

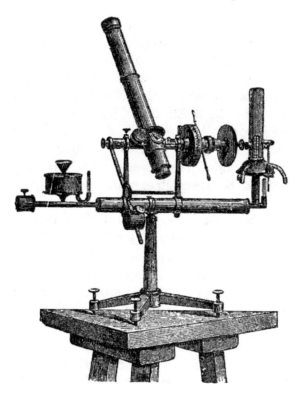

Fig. 2.6 In the Zöllner-photometer the light of a flame in the Bunsen burner on the right is projected into the field of view to act as a comparison star. Between the flame and the telescope focus a rotatable polarizer is used, which can adjust the light of the comparison star. The structure on the left is a counterweight. See also Fig. 10.11). From: *Efron Encyclopedic Dictionary* [29]

images of stars. For this he used a heliometer, that is a telescope in which the objective lens is cut in half and both halves can move independently. In this way he captured two random stars at the same time, which he de-focused, and then compared the brightness of the two blurred images. Such a telescope is called a heliometer, because it was originally developed to measure the diameter of the Sun by making the two opposite edges coincide in the image. The Zöllner photometer was a very significant improvement on previous methods and revolutionized the measurement of the brightness of stars. Zöllner photometers have been in use for decades.

There is no doubt that Kapteyn has used such an instrument. Indeed, the Utrecht Observatory had such a photometer, although preserved catalogs of the available instruments do not go back long enough to check that it had one in Kapteyn's time. But most likely it did. This also means that Kapteyn must have spent a lot of time on practical astronomy.

Proposition XV reads: *The average of the proper motions of stars of different apparent brightnesses is not inversely proportional to their distance.* This is fascinating; Kapteyn spent his whole life working on the distribution of the distances of stars, and this shows that he had already thought about this in his student days. In the next chapter we will see that William Herschel (1738–1822)—John Herschel's father—had assumed that all stars were intrinsically equally bright. Then their apparent brightnesses in the sky would be an indication of their distances. But stars are not at all equally bright intrinsically and have a wide range of 'luminosities'. The proper motion is the slow change of the position of a star in the sky as a result of its motion through space. If stars on average would move at the same speed, then the proper motion would be an indication, statistically at least, of their distances. The smaller the proper motion, the further away (grosso modo) the star is from us. Now, what Kapteyn was stating is that in fact it is more complicated. He must have researched the astronomical literature on this subject and must have already thought about the distribution of the stars in space, the subject to which he devoted his astronomical life.

So we see that for Kapteyn in Utrecht there was no opportunity to write a PhD thesis in astronomy, because at just this time the professor's chair in that disciple had been vacant. But he nevertheless developed a great interest in astronomy and must even have done extensive observational and theoretical work in it. In his thesis he expressed his gratitude for his training to Grinwis and Buys Ballot as his teachers; although Hoek was still a professor in the first years of his studies, Kapteyn apparently did not benefit much from him and did not think it was necessary to thank him.

2.4   Catherina Elisabeth Kalshoven

During the last years of his studies in Utrecht, Kapteyn became acquainted with the Kalshoven family and his future wife Catharina Elisabeth (Elise) (1879–1945). He came to know her probably through Elise's half-sister Jacqueline Kalshoven (1846–1926). She was married to Simon Brouwer (1833–1920), a notary who probably knew Kapteyn from the student fraternity. The Brouwer family lived in Maarssen, about 10 km to the north-west of Utrecht, and the Kalshovens lived there as well. Elise's mother, Henriëtte Mariëtte Augustine Albertine Kalshoven–Frieseman (1822-1895), was married to Jacobus Wilhelmus Kalshoven (1813–1869) in 1853 and had two daughters and a son with him; Elise was the oldest. Kalshoven was a manufacturer and seller of pianos; he had previously been married to Catharina Elisabeth Brandt (1814–1845) and had five children besides Jacqueline. After he was widowed he married Henriëtte Frieseman, but he had died by the time when Kapteyn got to know the family. Elise was named after Jacobus Kalshoven's first wife. It sounds a little strange to our modern ears—or maybe even inappropriate—that a father names his first child with his *second* wife after his *first* wife. Maybe to make up for this Henriette Hertzsprung–Kapteyn is named after her grandmother, Elise's actual mother.

Kapteyn was very much attracted by the atmosphere in the Kalshoven home, which was completely different from the environment in which he had grown up. In the words of Henriette Hertzsprung–Kapteyn [1]:

> One could say that were antipodes. The two young daughters of the house did not have much household duties; they were not studying or working, they were cheerful and sociable, they enjoyed life there in a simple and unsophisticated way.

Elise loved to play the piano accompanying her sister Marie Gabriëlle's (1857–1940) singing.

> Because the girls had never attended a school and had gained the little knowledge they had from a governess, they had retained a unique originality, which made for a special charm that immediately caught everyone's eye.

But although they went along well, the friendship between Jacobus and Elise did not yet develop into a courtship or an engagement.

3

Astronomy Around 1875

O, telescope, instrument of much knowledge,
*more precious than any scepter, is not he who holds thee
in his hand made king and lord of the works of God?*
Johannes Kepler (1571–1630)

*I have looked further into space than ever a human being
did before me. I have observed stars of which the light,
it can be proved, must take two million years to reach the Earth.*
William Herschel [1].

3.1   Astronomy of the Solar System

A little bit of astronomical background is needed for a good understanding of Kapteyn's work and appreciation of its significance. I start with the Solar System.

Our Solar System consists of eight planets, from the Sun first the 'terrestrial' planets with a solid surface, Mercury, Venus, Earth and Mars, and further out the 'gas giants' Jupiter, Saturn, Uranus[1] and Neptune. The properties of their orbits were described around the beginning of the seventeenth century by Johannes Kepler (1571–1630) in the form of his three laws, which specify that they are elliptical with the Sun in a focus and with a

---

[1]Uranus should be pronounced with the emphasis on the first syllable 'U'.

© The Editor(s) (if applicable) and The Author(s), under exclusive license
to Springer Nature Switzerland AG 2021
P. C. van der Kruit, *Pioneer of Galactic Astronomy: A Biography of Jacobus C. Kapteyn*,
Springer Biographies, https://doi.org/10.1007/978-3-030-55423-1_3

higher orbital velocity when the planet is closer to the Sun and a longer orbital period the further they are from the Sun. Subsequently, Isaac Newton (1642–1727) had demonstrated that these properties and laws could be derived directly from his theory of gravitation. Until the middle of the nineteenth century, great progress was made in the study of the disturbances of the orbits of the planets by their mutual attraction, which made it possible to calculate these effects with great accuracy. The high point of this early modern astronomy was undoubtedly the discovery of the planet Neptune in 1846.

Since antiquity, only the planets that are visible to the naked eye were known, Mercury up to Saturn. In 1781 William Herschel had accidentally discovered the planet Uranus during his systematic 'sweeping' (surveying) of the sky. He was a musician who had deserted from the army fleeing from Hanover to England, and had developed into a very successful amateur astronomer. The orbit of the new planet could be accurately determined, but even then deviations were found. This could only be due to the attraction of a planet even further away. The further away a planet is from the Sun, the longer it takes to complete an orbit around the Sun. Uranus then would overtake a planet further out regularly and during that passage the two would be relatively close to each other; the disturbance of the orbit of one planet by the other would then be the greatest. Indeed, we now know that Uranus has 'overtaken' that eighth planet, Neptune, at the beginning of the nineteenth century; around 1821/22 they were closest to each other, when they were on one line as seen from the Sun. Now it should be possible to calculate from the disturbances of the orbit of Uranus what the orbit of that disturbing planet would have to be, how massive that planet would be and where it would be at any given moment in its orbit. No easy task but not impossible although taking much time.

The Frenchman Urbain Jean Joseph le Verrier (1811–1877) had performed such calculations and had arrived at a prediction of the position of Neptune in the sky at particular times. Next Johann Gottfried Galle (1812–1910) found Neptune in Berlin in September 1846 at the position predicted by le Verrier. The Englishman John Couch Adams (1819–1892) had also made such calculations (also correctly and predicted a position for Neptune even somewhat earlier), but astronomers of the Royal Observatory in Greenwich had too little faith in it to check it out. It is ironic that Galileo Galilei (1564–1642) has actually seen Neptune in his field of view in 1611 during his first telescopic observations, when he discovered the four large satellites of Jupiter, because Neptune happened to be extremely close to Jupiter in the sky at that time. However, he did not recognize the planet as such, undoubtedly because of his primitive optics. Even more ironic is that John Frederick William Herschel (1792–1871), the son of William Herschel, observed Neptune in 1830, according to his notes, but also failed to recognize it as a planet in his telescope, despite the better quality of his images. So we very closely missed the interesting situation that the only two planets in our Solar System that are not visible to the naked eye would have been, both accidentally, discovered by a father and son.

On the first day of the nineteenth century, January 1, 1801, the first of a series of small planets or asteroids was observed, which we now know to be a swarm of objects, situated between the orbits of Mars and Jupiter. In 1802, 1803 and 1807 three more were found, and after a long interlude from 1845 onward even more of them. At the end of 1850 the number had risen to thirteen; at the end of 1870 the number was 157 and at the close of the nineteenth century, on 31 December 1900, it was 463. From a certain apparent regularity in the distances between the planets and the Sun one could expect that in the relatively large space between Mars and Jupiter there should be another planet. Planets form because dust particles fuse together to form smaller debris, called planetesimals. This then produced eventually the planets, but the gravitation of the largest planet Jupiter, which is close to the asteroid belt, must have disturbed this process.

In the nineteenth century people were already familiar with comets. These are objects that move in highly elliptical orbits. They can therefore get far from the Sun, at which time they are not visible for us; but they must return regularly. Edmond Halley (1656–1742) suggested after noting the close similarity in their orbits that the comet visible in 1682 may have been the same as that of 1607 and 1531, and predicted that it would return again in 1758/9. And several scientists such as Immanuel Kant (1724–1804) had already concluded that comets should consist of volatile materials that evaporated because of sunlight, which causes the spectacular tails of comets when they come close to the Sun.

From the positions in the sky of objects in the Solar System, it is possible to calculate with great precision the orbits of these objects relative to each other in space; these were therefore also known in the nineteenth century. But because only positions and no distances could be measured, the result was a good map of the Solar System, on which the orbits of the celestial bodies are accurately represented relative to each other, but the scale of that map would remained unknown. For this it is necessary to measure the distance between one of these objects and the Earth at a certain point in time. At first astronomers used the planet Mars for that, the first planet outside the orbit of the Earth. when Mars and the Earth are on one line with the

Sun, the distance between them is the smallest (i.e. only one third of the distance from the Sun to the Earth); that would be a good time to measure the distance of Mars. This could be done by measuring the exact position of Mars relative to the stars in the background, from two places on Earth as far apart as possible. Those positions will be different (the so-called parallactic effect); the difference depends on how far apart the two measuring points on Earth are (which distance is known) and on the distance to the Earth of Mars at that moment. The latter can then be calculated, and with that the scale of the Solar System is determined. This is usually expressed in the average distance from the Earth to the Sun, the 'Astronomical Unit'. This is 149.6 million kilometers. The angle at which the radius of the Earth would be seen from the Sun is of course also a measure for the distance from the Earth to the Sun. This angle is called the solar parallax and is 8″.79 (8.79 seconds of arc). In the middle of the nineteenth century this distance was only known to no better than a few percent.

There is a second method for determining the scale of our Solar System. For this you can use Venus (but in principle also Mercury). These two planets can stand between the Earth and the Sun and sometimes cross in front of the Sun. The exact time at which the planet passes the Sun's edge differs for different places on Earth, and can therefore also serve as a basis for a distance calculation. This method was first applied by Edmond Halley in 1676, using a transit of Mercury; however, the result was unsatisfactory because this planet is too far from the Earth. Venus transits provide much more precise observations, but they are rare. They occur in pairs with an interval of a few years, but in between those pairs the interval is about 120 years. The transits of 1761 and 1769 had already yielded a reasonable result. Various places on Earth were visited to observe the following transits of 1874 and 1882.

These Venus transits were extensively studied by the Canadian-American astronomer Simon Newcomb (1835–1909), who we will encounter below in relation to Kapteyn. He determined the solar parallax (and thus also the Astronomical Unit) to an accuracy of unprecedented 0.2%! In the twentieth century, the first method (the one with Mars) was successfully replaced by using asteroids, when some of these were discovered that can come much closer to the Earth than Mars.

In 1875 there was no understanding of the physical processes in the Sun that cause it to radiate. Of course it was possible to measure how much energy was involved. One could also calculate how much energy the Sun contains in gravitation, i.e. how much energy would be released if the Sun would contract to a very small size. From this one could deduce how long the

Sun could continue to shine at the rate it does. This calculation, made by William Thomson, 1st Baron Kelvin (1824–1907) and Hermann Ludwig Ferdinand von Helmholtz (1821–1894) and therefore named the Kelvin-Helmholtz contraction time, yields 30 million years. So, on that basis, it was assumed that the Sun could be at most millions of years old. That was much older than would follow from the dating of the Creation at 4004BC by bishop James Ussher (or Usher; 1581-1656) on the basis of Biblical chronologies.

3.2 Higher Parts of the Heavens

In 1875 people had a very good idea of the structure of the Solar System. Seen from Earth, the Sun describes an orbit in the sky, completing a full circle in a year's time. This orbit in the sky is called the ecliptic, and the constellations through which the Sun moves (if you could see them at that moment) are those of the zodiac, which we know from astrology. The orbits of the planets and asteroids are all close to the orbital plane of the Earth, so in the sky they are always quite close to the ecliptic. The study of the 'fixed stars' was much less developed than that of the planets. Because the stars can stand anywhere in the sky and do not care about the ecliptic, Kapteyn would later refer to this as the 'Higher Parts of Heaven' (in his speech as outgoing Rector Magnificus in 1891).

Most of the astronomers' research effort around 1875 went to the Planetary System. An inventory of scientific publications[2] from that period shows the following. About 35% concerned the orbits of objects in the Solar System and 25% the studies of meteors, the Sun and Moon (including eclipses) and the physical properties of planets. Nearly 20% was about instrumentation, photography, observational and mathematical techniques and astronomy related sciences. Only about 20% of the literature was devoted to the stars and nebulae outside our planetary system. At the time that Kapteyn entered university, almost nothing was known about the physical properties of fixed stars, nor about the origin of the light they emit, or about their formation and evolution. But since antiquity there had been catalogs describing their positions in the sky and their apparent brightness. Before I say more about this, I have to explain first how astronomers catalog the positions of stars and quantify brightness.

We represent the sky as a sphere that we look at from the inside. The position of a star in the sky is measured in the same way as the geographical

---

[2]See Sect. 3.8 and Table 3.1 in my *Jacobus Cornelius Kapteyn; Born investigator of the Heavens* [5].

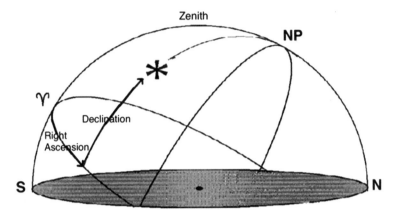

Fig. 3.1 The coordinates of a star in the northern hemisphere. For details see the text. Figure by the author

longitude and latitude on Earth, namely relative to two points. On Earth these are the North Pole and the point on the equator that lies on the Greenwich meridian. The situation is presented in Fig. 3.1. The shaded surface is that of the Earth and the half sphere represents the sky. The intersection line between sphere and plane is the horizon with the directions north and south marked with 'N' and 'S'. The observer is in the center of the spherical dome. The point in the sky right above the observer is called the zenith. The circle from south on the horizon via the zenith to north is called the meridian. The straightforward way to represent the position of a star is as in the left-hand part of Fig. 3.2 with the azimuth along the horizon (usually measured from

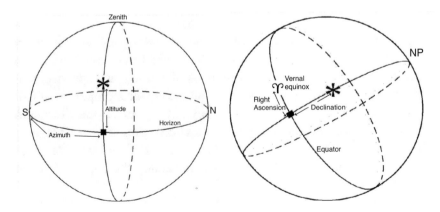

Fig. 3.2 Coordinates on the sky of a star in azimuth and altitude (left) and in right ascension and declination (right). The symbol ♈ indicates the vernal equinox. Figure by the author

the south point 'S') and the *altitude* the height above the horizon. This of course changes all the time, and is not useful for a catalogue, but important to set a telescope at a given time from a given location on an astronomical object.

Therefore astronomers use a coordinate system that rotates along with the sky, i.e. relative to the poles and the equator 90° away from these (right-hand panel in Fig. 3.2). One of the two coordinates is comparable to the geographical latitude and is called *Declination*; this is the angle between the star and the projection on the equator. So it is also 90° minus the angle from the star to the North Pole (South Pole in the southern hemisphere; the declination is then negative). The other coordinate, equivalent to the geographical longitude is called it Right Ascension (in Fig. 3.1 indicated by R.A.). This is the angle from the projection of the position of that star on the equator to the point (the 'vernal equinox') where the Sun crosses the equator in the northern spring (i.e., on 20 or 21 March). This is indicated in Fig. 3.1 with the symbol '♈'.

Figure 3.1 shows the situation with respect to the horizon and zenith. In the sky, the North Pole ('NP') is the point directly above the Earth's North Pole. As seen from any position on Earth, the sky rotates around its axis once a day; the celestial poles stand still. In the northern hemisphere, the star Polaris happens to be close to the North Pole (currently about two-thirds of a degree from it). Figure 3.3 shows the diurnal rotation on a long exposure. The position of the pole above the horizon can be determined with circumpolar stars, which are so close to the pole that they do not set. One then measures the height above the horizon when such a star passes through the meridian above and below the pole; the pole is then exactly midway between the two measurements.

In Fig. 3.1 the time is chosen such that the vernal equinox just crosses the meridian. The sky rotates and the vernal equinox, like the stars, rises every day in the east and sets in the west. Accurate measurement of the position of the Sun is necessary to know where this vernal equinox is located relative to the stars at any given moment.

The easiest way to determine these coordinates is to take (in the northern hemisphere, but for the southern hemisphere the procedure is similar) the moment at which the star passes through the meridian. One then measures the angle above the horizon and the exact time. The declination follows from the first measurement, because it is known how far above the horizon the North Pole is at one's location of observation. From the second, that of the exact time of the meridian passage of the star, you deduce the right ascension. Because the latter is a time measurement, right ascension is

Fig. 3.3 The daily rotation of the celestial sphere. This photograph was made at the Paranal Observatory of the European Southern Observatory ESO in Chile. On the right we see the building of one of the four 8.2 m unit-telescopes of the Very Large Telescope VLT. Each star traces an arc that is part of a circle around (in this case) the South Pole in the sky, in the center of the picture. With a 12-h exposure, that would be half-circles. This photograph has been produced with an exposure time of about an hour. The faint broad swipe from the center left to the top right is the Milky Way and the blurs to the left of the telescope are the Magellanic Clouds. European Southern Observatory/Iztok Boncina [33]

usually not expressed in degrees but in hours (360° then becomes 24 h, and so on).

The apparent brightness of stars in the sky is indicated by magnitudes. In ancient times this was no more than a visual estimate, with the brightest stars by definition having magnitude 1 and the faintest, that the naked eye could see, magnitude 6. Later, in 1856, the British astronomer Norman Robert Pogson (1829–1891) formalized this scale by exactly equating a difference of 5 magnitudes to a factor of 100. Each magnitude then becomes a factor of 2.512, not too different from the ancient scale. In this definition, the very brightest stars became brighter than magnitude 1 (the brightest star Sirius has magnitude –1.5). The faintest stars that can be seen with the naked eye under the best conditions are still magnitude 6. Over the entire (northern plus southern) sky, the total number of stars to magnitude 6 is somewhat more than five thousand.

The most famous classical catalog is that of Hipparchus of Nicaea (c.190—c.120 BC). It was he who, as far as we know, first used the magnitude scale. He made a remarkable discovery, namely that the position of the zero point of the right ascension changes over time. This point, called vernal (spring) equinox because then day and night are equally long, changes because the rotation axis of the Earth does not stand still in space. Compare this to a spinning top. When it is inclined, the attraction of the Earth will want to pull it down, but that is kept in check by its rotation. The axis of the spinning top then rotates while maintaining the angle to the Earth's surface. The same goes for the Earth. The axis makes an angle of about 23° with a line perpendicular to its orbital plane and because the Earth is slightly flattened, the Sun applies a torque or rotational force to it, pulling the axis of the Earth towards that plane. But the rotation of the Earth, just like with the top, ensures that the rotation axis swings around, always maintaining the same angle with the orbital plane. This so-called precession involves a slow circular motion of the North and South Poles in the sky around the poles of the ecliptic; a complete 360-degree revolution, after which the poles return to the same point on the sky, takes about 26,000 years. As a result of the precession, the equinox shifts more than 50 seconds of arc per year, so the exact value of the equinox is of great importance. Right ascension and declination change slowly with time and you then have to allow for this when comparing the positions of a star at different times. The amount of the change of the equinox was not yet known very precisely in the middle of the nineteenth century; it was not until around 1900 that Simon Newcomb, whom we already met in connection with the solar parallax, determined a value that was accurate by modern standards.

Claudius Ptolemæus (Ptolemy) (c.90–c.168) used Hipparchus' catalog, supplemented with his own observations, to publish a new catalog in his famous book *Almagest*, which summarized the astronomy of his time. This catalog, which has been preserved, contains over a thousand stars with positions accurate to about 10 arc minutes (one third of the diameter of the full Moon in the sky).

Star catalogs became more and more important for navigation, and more accurate star positions were required. An early method to determine the positions of stars was using a quadrant or sextant (see Fig. 3.4). As with the legs of a compass, one looked along two lines at two different stars, or at a star and the horizon. On the scale along the quadrant (rectangular) or sextant (60°-construction) the angle was then read off. Later there were wall quadrants, which were often pointed at the meridian, i.e. in the direction to the south or north. The Dane Tycho Brahe (1546–1601) was a famous

Fig. 3.4 Positional measurement of stars, as done by Hevelius in the seventeenth century with a sextant [34].

cartographer of the stars. His unprecedentedly accurate measurements of the positions of the planets enabled Kepler to derive his laws for the planetary orbits. Johannes Hevelius (1611–1687), mayor of Danzig, produced a catalog with 1564 stars, which was published posthumously. The accuracy of the star positions had improved to 2 minutes of arc by Brahe and to about 20 seconds of arc by Hevelius.

One of the tasks of the Royal Greenwich Observatory when it was founded, was to perfect methods for determining positions at sea. The lack of accurate clocks for keeping track of the time to determine right ascension and thus geographical longitude, remained for a long time the greatest limitation (see *Time and time again* by my former student Richard de Grijs [36] for an excellent introduction to the subject). Especially in the seventeenth

Fig. 3.5 The southern constellation Southern Cross, the kite-like configuration, is on the right in this wide-angle view of the southern sky. The two bright stars on the lower left are α and β Centauri, which are called the 'pointers'. The dark blot in the Southern Cross can also be seen with the naked eye and is called the 'Coalsack'. It is interstellar dust that absorbs and scatters the light from stars behind it. European Southern Observatory/Yuri Beletsky [35]

century there was also the problem that the star catalogs were based on observations in the northern hemisphere; the southern sky had not been mapped with the same accuracy or completeness. This was of course required for navigation in southern oceans. Several attempts had already been made in the seventeenth century to correct this situation. Kepler used material from observers in the southern hemisphere. Already in 1603 Frederick de Houtman (1571–1627) published a catalog, based on two Dutch expeditions to the East-Indies, where he also introduced southern constellations (such as the Southern Cross, see Fig. 3.5). Edmond Halley was assistant at the Royal Greenwich Observatory (to John Flamsteed; see below) when he traveled to St. Helena in the South Atlantic Ocean in 1677/8 and made observations there for about a year. Frank Verbunt en Rob van Gent of the University of Utrecht recently published a detailed discussion and analysis of early southern star catalogs [37]

After Hevelius, only telescopes were used for observations of the sky. John Flamsteed (1646–1719) was the first 'Astronomer Royal' and founder of the Royal Greenwich Observatory. His (also posthumously) published star catalog contained 2935 stars with a positional accuracy of around 10 seconds of arc. James Bradley (1693–1762), third Astronomer Royal, improved the accuracy to just a few seconds of arc. His 60,000 observations of 300 bright stars are remained unreduced (corrected for observational effects), but they were later used by Friedrich Wilhelm Bessel (1784–1846) to published a catalog in 1818. The state of the art around 1875 was the work of Friedrich Wilhelm August Argelander (1799–1875) from the Bonner Sternwarte. He undertook the *Bonner Durchmusterung (BD)*, for which he observed a total of some 325,000 stars between 1852 and 1862. He determined their positions with an accuracy of one-tenth of a minute of arc and brightnesses to within 0.1 magnitude. This was a complete sample up to about magnitude 9, but it also contained many stars of magnitude 10. But again, it concerned only stars in the northern hemisphere. Under his successor Eduard Schönfeld (1828–1891) an extension to the south (but from Europe) was made between 1875 and 1881 which went as far as southern stars can still be seen from Bonn (the *Südliche Durchmusterung*, with 133,000 stars). The plan was then to further measure the southern sky from the Córdoba Observatorio Astronómico in Argentina, but that only started in 1892. In the meantime, the use of meridian circles (see Fig. 3.6), telescopes that can only move around one axis and can only observe in the meridian, had become common practice.

Stars do not stand still with respect to each other in the sky, although over a lifetime the change is imperceptible with a few exceptions. Already in 1718 Edmond Halley noticed that the bright stars Sirius, Arcturus and Aldebaran had changed their position by more than half a degree compared to the catalog of Hipparchus. As we already saw, this motion in the sky (usually expressed in seconds of arc per year or per century) is called the 'proper motion'. The fact that the Sun does not stand still in space either, is shown by the systematics of the proper motions of stars across the sky. Think of a straight road in a landscape in which two trees stand in the distance on either side of the road. When you drive a car on that road towards the trees, you see them first close together, but then their separation increases faster and faster until they are opposite each other at the moment you pass them. Eventually, they will get back together behind you. If all stars except the Sun would stand still, the proper motions of all stars in the sky would be directed towards the point where the Sun (and the Earth) comes from, and would move out from the point where the Sun moves towards (the Apex).

Of course, this pattern is less clear in the case of space motions by the stars themselves, but it would still be possible to find it as a systematic pattern superimposed on a presumably random distribution of proper motions. This was confirmed by William Herschel as early as 1783 on the basis of

Fig. 3.6 Positional measurements of stars were performed with meridian circles. This is the one of the Sterrewacht Leiden [38]

the measured proper motions of at first 12, later 36 stars. The position of the Apex, which he derived, corresponds remarkably well to modern determinations. The Sun moves in the direction of the constellation Hercules (about 15° from the bright star Vega in the constellation Lyra). The speed of the Sun relative to the stars in its neighborhood can only be determined (statistically, if we have large numbers of stars) if we know the radial velocity of those stars. Such radial velocities of stars can be measured using the effect that light or sound gets a higher frequency (or smaller wavelength) when an object moves towards us, and vice versa. This effect was discovered in 1842 by Andreas Doppler (1803–1853). The velocity of a star follows from the shift of spectral lines: dark lines of known stationary wavelength in the spectrum. Spectroscopy of stars, however, was only possible towards the end of the nineteenth century, and then only for the brightest stars. Then it became clear then that the velocity of the Sun relative to the collection of stars in the the solar neighborhood is 17–18 km/s.

Another concept that we will encounter frequently is that of parallax; I refer to Appendix A.2 for a more technical discussion. Ever since the introduction of the Copernican or heliocentric model of the Planetary System, the aim had been to observe the reflection the motion of the Earth around the Sun in the positions of stars in the sky. Each star will describe an ellipse in the course of the year as a result of the changing direction of observation from the moving Earth. The further away the star, the smaller the ellipse. For stars that are relatively close by, this motion should be measurable by comparing the position of a nearby star to stars in the background that are further away. The semi-major axis of the ellipse is called the 'parallax' and that is thus also a measure for the distance.

Now, despite detailed attempts, parallaxes could not be measured and so the heliocentric model was either incorrect, or the distances of the stars were too large. The latter, of course, is the case. Finally Friedrich Bessel succeeded in 1838 in measuring the parallax of the star 61 Cygni in the constellation Cygnus (0.31 seconds of arc); Thomas James Henderson (1798–1844) from Scotland did the same in 1839 for α Centauri. The latter—in the southern constellation Centaur, see Fig. 3.5—is the closest star to our Solar System. It is at 4.35 lightyears away from us, where a lightyears is the distance that the light travels in a year. At 300,000 km/s, a lightyear comes out as 9.5 trillion (9.5×10^{12}) km. Astronomers usually use the unit parsec (pc); that is the distance of a star with a parallax of 1 second of arc, and that amounts to 3.26 lightyears. The parallax of α Centauri, which is at 1.3 pc, is therefore 0.75 seconds of arc (indicated as $0\rlap{.}''75$) and that is the largest parallax of all stars (except for the Sun of course). We now know, thanks to accurate parallax measurements from space by the ESA satellite Hipparcos, that there are only 271 stars with a parallax greater than $0\rlap{.}''1$; most of them are so faint that in the middle of the nineteenth century astronomers had not even attempted to measure their parallaxes.

3.3   The Construction of the Heavens

It was the aforementioned William Herschel who was the first to attack the problem of the spatial distribution of the stars. He called this the Construction of the Heavens. He built his own telescopes, the largest and best of his time, using mirrors of speculum, an alloy of copper and tin in a ratio of two to one, which he himself ground into the desired shape. He did this at his house in Slough, over 30 km west of London and not far from Bath. His largest telescope was the so-called 40-foot telescope (see Fig. 3.7). This 40 ft (just over 12 m) refers to the focal length of the primary mirror; this mirror had a diameter of 48 inch. (1.20 m). That was huge and unprecedented at the time. He carried out his first study with a predecessor; that was a 20-foot telescope, which looked a lot like the 40-foot telescope depicted. He kept

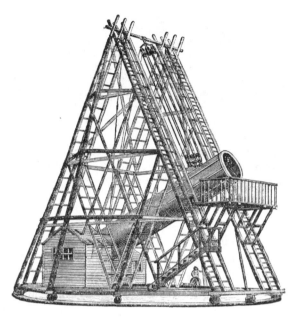

Fig. 3.7 The large 40-foot telescope by William Herschel. The scale can be appreciated by comparing to the human figure in the background. The diameter of the tube is about 1.20 m. The 40 ft (some 12 m) refers to the focal distance. Wikimedia Commons [39]

the telescope steady and counted the stars as they drifted through his field of view as a result of the rotation of the sky (of the Earth actually). His sister Caroline Lucretia Herschel(1750–1848) sat by an open window and jotted down the counts he shouted at her.

For his cross-section of the 'Sidereal System' (see Fig. 3.8) he apparently counted stars to about magnitude 15, along a circle in the sky more or less perpendicular to the Milky Way (see my contribution to the *Legacy* symposium *The Legacy of J.C. Kapteyn* by myself and Klaas van Berkel [4] for details in how this was derived). He assumed that his telescope could see as far as the edge of the Galaxy and that the stars were uniformly distributed along his lines of sight. From the counts one then can calculate immediately the distance to the edge (in some unit). With his later 40-foot telescope it turned out that even more fainter stars were visible than with his smaller telescope, and his assumptions turned out to be incorrect. He also assumed that all the stars were intrinsically equally bright. From the relative positions of components of binary stars he deduced that the two components moved around each other (and therefore had to be at the same distance), while their apparent brightness often differed considerably. So they also had to have intrinsically different luminosities. Towards the end

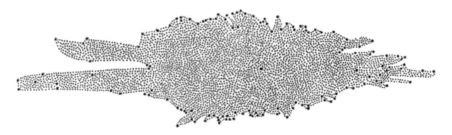

Fig. 3.8 The famous cross-section of Herschel's Sidereal System based on his 20-foot star counts (this is a smaller and earlier version of the 40-foot telescope in Fig. 3.7). The point in the middle is the position of the Sun. From *The Construction of the Heavens* (1785) [40]

of his life he was convinced that the Galaxy was flattened, as if it were bounded by two planes, between which it stretched infinitely far. Herschel was the first to attempt to derive an model of the spatial distribution of stars.

In summary, around the time that our main character, Jacobus Cornelius Kapteyn, graduated from university in 1875, a great deal of knowledge about the structure of the Solar System was already available to astronomers, but the astronomy of the stars was still in the cataloging stage. Nothing was known about the physical characteristics of stars, nor did astronomers know anything about their structure, formation and evolution. Little could be said about their spatial distribution; especially this last problem of the structure of the Sidereal System would fascinate and occupy Kapteyn throughout his life.

Fig. 4.0 Kapteyn in the early seventies of the nineteenth century when he was a student in Utrecht. This illustration comes from Henriette Hertzsprung–Kapteyn's biography [32]

4

Observator and Professor

I was born on January 8, 1942, exactly three hundred years
after the death of Galileo. I estimate, however,
that about two hundred thousand other babies were also born that day.
I don't know whether any of them was later interested in astronomy.
Stephen William Hawking (1942–2018)

It is no prudence, but a waste of the Kingdom's money
to appoint a professor without supplying him
with the indispensables for his teaching.
Jonkheer Bernard Hendrik Cornelis Karel van der Wijck (1836–1925)

Both Henriette Hertzsprung–Kapteyn in her biography [1] of her father Jacobus Kapteyn as well as various writers of obituaries or other articles about Kapteyn suggested that he had become an astronomer accidentally. He obtained his doctoral degree in Utrecht with a dissertation in mathematics; it is stated that he subsequently heard by chance that a position as observator at the Sterrewacht Leiden had become vacant, to which he applied successfully. But this assumption is not consistent with what we know of his great interest in astronomy, as can be seen, for example, from the star map he made as a teenager and the astronomical propositions presented for his PhD thesis defense.

Bernhard Jonkheer (Squire) Hendrik van der Wijck was professor of philosophy and literature at the University of Groningen. This quote is from his lecture on the occasion of the end of his term as Rector Magnificus in 1879 [1].

P. C. van der Kruit, *Pioneer of Galactic Astronomy: A Biography of Jacobus C. Kapteyn*, Springer Biographies, https://doi.org/10.1007/978-3-030-55423-1_4

Father Kapteyn had wanted his son Jacobus to accept a position as a teacher. An HBS was an excellent place to work for those young men that had obtained a PhD. We saw earlier that his older brother Willem was a teacher at the HBS in Middelburg when he was appointed professor in Utrecht. Such a job would be an obvious choice for Jacobus as well.

What about military service, though? After all, all the healthy young men had to fulfill military service. The archives of the municipality of Barneveld show that Kapteyn had passed the medical, physiological and mental inspection. The young men that would have to report for active service were determined by drawing lots. Kapteyn was the fifth draw and would normally have had to go and serve in the army, but that he had bought off by hiring a so-called 'remplaçant'. This replacement was a man who had also passed, but in one way or another was exempt from military service (e.g. when two older brothers had already been in active service, or when he was a higher draw when sufficient young men had been chosen). Hiring a remplaçant was reasonably expensive; it cost like 600 to 800 guilders (about two years' income of an unskilled worker), of which a large part could be absorbed by mediators and civil-law notaries. Only the reasonably wealthy could afford it.

The records in the Barneveld Archives on the selection also give us insight into the height of Kapteyn. He was recorded as being 171.5 cm. For the 54 men approved in Barneveld in the same year as Kapteyn, the median was 167.2 cm, so Kapteyn was on the tall side, although not extremely so.

4.1 The Sterrewacht Leiden

Astronomy had been practiced in Leiden, as mentioned above, since shortly after the founding of the university. In 1632 Jacobus Golius (1596–1667) bought a quadrant, which was built by Willibrord Snellius (1580–1626). Snellius used this instrument to measure from one church tower what the angle was between two other church towers in the distance. If you know the distance between the two latter towers, you can calculate how far away you are from the first. By moving on to other towers, you can finally measure the distance between two fairly distant cities, which are chosen to be more or less on a north-south axis (in this case Alkmaar and Breda, about 116 km apart, see Fig. 1.0). If you then measure from these two places how high a certain star culminates above the horizon (passes the meridian in the south), you can calculate from the difference between those angles what the difference in geographical latitude is and thus determine the circumference of the Earth. But you can also use this instrument to determine angles between stars in the sky. The quadrant of

Fig. 4.1 The Sterrewacht Leiden as it was originally depicted on the cover of the Annals, in which scientific results were published. Archives Sterrewacht Leiden

Snellius was installed on a platform on the top of the Academy building. Until the nineteenth century, astronomical observations were made from there.

In 1861, the 'Sterrewacht', a real observatory with offices, domes and telescopes, was inaugurated in the university's hortus botanicus (see Figs. 1.3 and 4.1). The director and professor of astronomy Frederik Kaiser (Fig. 4.2) had modeled it after the Pulkovo Observatorium near St. Petersburg. He had built up quite a reputation for highly accurate positional measurements of stars and other objects (such as asteroids). His successor Hendricus van de Sande Bakhuyzen, who took over the directorship in 1872 (see Fig. 4.3), further developed this reputation. For observing stars and determining their positions with the meridian circle (see Fig. 3.6), he was assisted by two observators. At the beginning of 1875 these were the German astronomer Karl Wilhelm Valentiner (1845–1931) and Ernst Frederik (1848–1918), the younger brother van de Sande Bakhuyzen. In the course of that year, however, Valentiner was appointed director of the Mannheim Sternwarte and on the recommendation of Hendricus van de Sande Bakhuyzen, Kapteyn was eventually appointed by the Minister of Internal Affairs (who was in charge of the universities in the country) as second observator. It is interesting to examine the proceedings of Kapteyn's appointment as successor to Valentiner in more detail.

Fig. 4.2 Frederik Kaiser, director of the Sterrewacht Leiden, under whose directorship the building was realized. Archives Sterrewacht Leiden

From the documents of the curators of Leiden University we know that Hendricus van de Sande Bakhuyzen informed them of the intended departure of Valentiner in June 1875; they in turn informed the Minister. The latter indicated that the appointment of a successor would follow as soon as possible after receipt of a proposal. Van de Sande Bakhuyzen did submit a proposal on August 10 to appoint Kapteyn as second observator, in addition to the honorable resignation of Valentiner and the appointment of his brother Ernst as first observator. He mentioned that Kapteyn, at his request, had already worked at the Sterrewacht for some time, from which his suitability for the position had been confirmed. He hinted that there were possible other candidates, but that Kapteyn was to be preferred. It then remained quiet for some time.

On September 21, Hendricus van de Sande Bakhuyzen received a letter from the Minister suggesting another possible candidate. 'Curators have recently nominated Dr Kapteyn for the position of observator. The Minister suggested however that Dr Gleuns should also be considered. [...] You know him from his numerical work at the Sterrewacht.' He was requested to provide arguments for his choice. At the end of the letter the Minister wrote: 'And then also clarify who this Dr Kapteyn is + what he has done so far. Curators

Fig. 4.3 Hendricus Gerardus van de Sande Bakhuyzen. From Wikimedia Commons [42]

have not spent a single word on this, which is a nomination like this one is a bit meagre.' It looks as if the minister had a clear preference for Gleuns. Now Willem Gleuns (1841–1906) had studied mathematics in Groningen and had indeed spent some time at the Sterrewacht at his own request and had done some mathematical work there. Probably someone had exerted pressure on the Minister, for example Gleuns' father, Willem Gleuns Jr. (1808–1881), who also had studied in Groningen and in 1837 had obtained a doctorate on a 'mathematical-astronomical' study on sunspots. His supervisor was the professor of mathematics, who also was responsible for teaching astronomy, Jan Willem Ermerins (1798–1869). Gleuns, the father, was a teacher of mathematics at the HBS in Groningen and author of several textbooks and articles, including on astronomical subjects, and enjoyed some national fame.

Van de Sande Bakhuyzen hurried to answer; the next day already he sent an extensive letter, which I located with some difficulty in the National Archives. He explained in extenso that Dr. Gleuns had hardly any practical experience in astronomy or with telescopes and had shown no evidence of a special interest in astronomy. The calculations he had done at the Sterrewacht did not require any real knowledge of astronomy. He was in Leiden to do research in the field of the history of mathematics and in that context he also did some historical studies in astronomy, in which the library of the Sterrewacht came in handy. Van de Sande Bakhuyzen considered him to be very suitable for education, but absolutely not as observator. Of the four pages of the letter, two and a half are devoted to Gleuns. That part ends with: 'As an observer, however, I would not be able to work with him'. About Kapteyn Hendricus van de Sande Bakhuyzen wrote:

> Opposite to Dr Gleuns we have Dr Kapteyn. During his studies in Utrecht, he spent as much time as possible at astronomy and actually observed at the Sterrewacht there with its instruments, so that he was well trained in the use of small astronomical tools. Last year he planned to continue his studies at the Sterrewacht Leiden, but financial objections prevented him from doing so. Mr. Kapteyn remained in Utrecht therefore and defended his dissertation there in June of this year on a very good PhD thesis. When Dr Valentiner's departure created a vacancy I immediately thought of Mr Kapteyn as one of the candidates for the position of observator, and the very favorable recommendations, which I received about him from his professors Buys Ballot and Grinwis led me to invite Mr Kapteyn to come and work at my Sterrewacht for a while, in order to be able to decide if he were suitable for the position of observator. Mr. Kapteyn accepted this offer and has now worked here for three months, during which time he has very well lived up to the expectations that I initially held concerning him. Although he is, of course, not yet fully experienced in of all the observations that he will have to make here, he has nevertheless shown himself to be a very good observer; even these days he showed me an extensive series of observations, which was in no way inferior to those of many more experienced astronomers. Dr Kapteyn has already worked on several aspects of theoretical astronomy, while his considerable mathematical knowledge and skill will certainly enable him to make up for what he lacks in that respect soon.

On this basis he concluded that he could confidently recommend Kapteyn. After having received this letter, the official who dealt with the case at the Ministry wrote on it in pencil: 'For the Minister: So it will have to be J.C. Kapteyn.' From all of this becomes clear that Kapteyn did not end up an astronomer by chance, as has sometimes been stated in obituaries about him, but always had wanted to become an astronomer.

When Kapteyn came to Leiden a major maintenance of the instruments and buildings was on the way; in 1875 it was the meridian circle's turn. Kapteyn was to contribute to and help with this. Measuring positions of stars in the sky required extreme care to achieve the desired accuracy. To prevent distortion of the telescope by thermal expansion, it was even shielded from the heat of bodies: the observer himself and sometimes a colleague who made notes. Of great importance was the difference between the temperature of the air in the observation room and that outside. The roof was opened well on time around sunset, but improvements had to be made to the observation room itself. The director seized the opportunity to thoroughly refurbish the instrument itself; for that it had to be disassembled and sent to the manifacturer, J.G. Repsold and Sons in Hamburg. Before that could be done, everything had to measured and documented, from the precise positions of the marks on the reading scales to the characteristics of the threads in the positioning screws and the telescopic tube's stiffness. This was necessary in order to be able to make any corrections to the fifteen-year-old archive of observations. The observators did most of this work in practice, which was an excellent way for Kapteyn to gain experience in astrometry and positional astronomy. Also much work with regard to the determination of star positions continued. For example, there was another possibility to measure the solar parallax, when Mars in 1877 would be relatively close to the Earth during opposition. In opposition, a planet, Mars in this case, is opposite the Sun from the Earth's point of view and is therefore as close as possible to the Earth. The planet will then be visible all night long. The project required very accurate measurement of the relative positions of stars in the Mars environment in the sky. And that had to be done at many different observatories. At the Sterrewacht Leiden the meridian circle could not be used, so other telescopes were used for this purpose.

In Leiden, Kapteyn made some friends for life. He had regular dinners with a group of contemporaries, including the later famous botanist William Burck (1848–1910) and Ambrosius Arnold Willem Hubrecht (1853–1915). Hubrecht later became professor of zoology and anatomy in Utrecht and was well-known for his work in embryology. He was a son of the Secretary General of the Ministry of the Interior, which covered the universities, and therefore he sometimes had some inside information. One evening Hubrecht ordered a bottle of champagne and made a toast to Kapteyn's brother, Willem, who had been named a professor. This to Kapteyn's surprise. And then Hubrecht ordered a second bottle to make a toast to Kapteyn himself, who had been appointed professor in Groningen. Both appointments had not even been announced officially yet, but this happened shortly after in the Staatscourant of December 14, 1877 (see Fig. 1).

4.2   Professor in Groningen

As I described in Chap. 1, Kapteyn's chair was specially created as a result of a new law on higher education. The University of Groningen did not have a dedicated professor of astronomy before then; astronomy was taught starting with the founding of the university in 1614 by the States of 'Stad en Ommelanden' ('City and Surrounding Lands', now more or less the province of Groningen). Ubbo Emmius (1547–1625), who was the first Rector Magnificus, had in that year appointed Nicolaus Mulerius (1564–1630) as professor of medicine and Greek, but soon he was designated to teach mathematics, which also covered astronomy. Mulerius is best known for his annotated reissue of the book *De Revolutionibus Orbium Coelestium*, in which Nicolaus Copernicus (1473–1543) set out his heliocentric worldview. This is remarkable. This period was dominated by the religious disputes on the issue of predestination between the Remonstrants (in Dutch know as 'Rekkelijken' or non-dogmatic ones), the followers of Jacobus Arminius (1555–1609), and the Counter-remonstrants ('Preciezen' or dogmatic ones), followers of Franciscus Gomarus (1563—1641). In this the University of Groningen had usually chosen the side of the Counter-remonstrants. Mulerius was a strict Calvinist as well and—at least to the outside world—a supporter of the old geocentric worldview. His painting in the Senate Room in the Groningen Academy Building (Fig. 4.4) shows him with two instruments in his hands to demonstrate spherical astronomy, the description of positions and orbits on the celestial sphere. After Mulerius astronomy had always been taught, sometimes by professors who had much interest in it, sometimes by others who considered it simply a task without any special appeal. There had been talk at some time of building an observatory, but the necessary funds were never found. Astronomy courses did not get beyond practical lessons on the roof of the Academy building (see Fig. 4.5), sometimes with the help of a small telescope (that was privately owned!).

After the introduction of the Higher Education Act of 1876, the Faculty of Mathematics and Natural Sciences regarded the appointment of a professor of astronomy as its highest priority, although the professor would also have to teach other subjects, such as statistics, meteorology and theoretical mechanics. Rudolf Adriaan Mees (1844–1886), professor of physics (see Fig. 4.6), but who also taught astronomy, needed urgently to be relieved of some of his tasks. The Faculty sent a nomination for an astronomy professor to the curators on November 19, 1877. There were five candidates (soliciting applications was not done at that time). From these, Kapteyn was the first choice and the second possibility was Charles Matthieu Schols (1849–1897). Schols was an engineer

Fig. 4.4 Nicolaus Mulerius (1564–1630), who was the first to teach astronomy in Groningen. University Museum Groningen [43]

and was very knowledgeable in mechanics, but not primarily an astronomer. It is possible that he was chosen in second place for tactical reasons, in case the curators or the minister would not agree with the first choice. The other three were astronomers and this would prevent the final choice to be one of those. Among these was Valentiner, Kapteyn's predecessor as observator in Leiden, and the other two were Ernst Emil Hugo Becker (1843–1912) of the Babelsberg Sternwarte near Berlin and (Ritter) Hugo Hans von Seeliger (1849–1924), whom we will meet again later. He had worked at the Bonner Sternwarte until 1873 and now taught astronomy at the University of Leipzig.

The curators did not accept the nomination completely and at the end of December sent a nomination to the minister to appoint Kapteyn as professor of astronomy, statistics and theoretical mechanics, with Becker in second place. This too may have been tactical, because the minister was unlikely to prefer a German to a Dutchman, especially if no observatory would be available, at least for the time being. It is very likely that Kapteyn's nomination in Groningen

Fig. 4.5 The Academy Building of the University of Groningen at the time Kapteyn was appointed to the position of professor. It had been in use since 1850, but would be completely destroyed in a fire in 1906. University Museum Groningen [44]

Fig. 4.6 Rudolf Adriaan Mees (1844–1886), professor of physics, who taught astronomy in Groningen until the arrival of Kapteyn. University Museum Groningen [45]

was on strong recommendation of both Hendricus van de Sande Bakhuyzen from Leiden and Jean Oudemans from Utrecht. The appointment followed by Royal Decree in December. For Kapteyn, this must have come as a complete surprise. Kapteyn accepted his appointment as professor on February 20, 1878 with an inaugural lecture, after taking an oath of loyalty to the King (Willem III) and his formal admittance to the Senate of the university (the collected professors). The text of Kapteyn's inaugural lecture has not been preserved; we only know the title from the university reports, namely *The annual parallaxes of the fixed stars*. Very appropriate, because distances and the spatial distribution of the stars were the subjects on which Kapteyn would continue to focus his research throughout his career. The distances of stars were also the subject of one of the propositions attached to his PhD thesis in Utrecht.

4.3   Without an Observatory

Kapteyn immediately went to work; to begin with, he attempted to obtain an observatory in Groningen. Already three days after his appointment, there was a meeting of the Senate, during which an request was approved for the provision in the budget application that was to go to the Minister. The list of desiderata was long, with ten items for the Faculty of Mathematics and Natural Sciences. Kapteyn's observatory was the third priority. The costs were estimated at 14,000 guilders, which in current purchasing power amounts to some 150,000 €. A physics laboratory (85,000 guilders) and a mineralogical museum and laboratory (5000 guilders) were higher on the list. The budget for the construction of an observatory, however, did not include the land and instruments, which would have to come separately from special subsidies. The following items were mentioned for the observatory:

	Guilders
heliometer	19,200
clock	1,000
chronometer	500
universal instrument	400
prism binocular with stand	200

We have encountered the heliometer before: this is a telescope of which the objective lens is cut in two, and where the two halves can move independently. In this way, two stars and their distance (the angle on the sky) can be measured from the amount of rotation between the two lens halves. A universal instrument is a small telescope that can be rotated around two axes, one horizontal and one vertical. This allows the position of a celestial body to be determined at any time (i.e. by measuring the angles azimuth, usually from the south along the horizon, and height above the horizon).

Eventually Kapteyn received 250 guilders (~2700 €) for the purchase of a simple telescope, which he had to advance from his private money, and 200 guilders for other instruments. By way of comparison, the annual salary of a starting professor like Kapteyn will have been about 4000 guilders. This was the first in a long series of frustrations about the lack of funds for an own observatory. It was also the reason for the remark of Rector Magnificus jonkheer Bernard H.C.K. van der Wijck, quoted at the beginning of this chapter, who at the time of the transfer of the rectorate in 1879 pointed out that at least 6000 guilders was necessary for an observatory for Kapteyn. In her father's biography Henriette Hertzsprung–Kapteyn [1] wrote:

> It never got any further, because no one supported him in his applications to the government. The other two astronomers, Prof. van de Sande Bakhuyzen, director of the Sterrewacht in Leiden and Prof. Oudemans in Utrecht, gave the Minister negative advice. They were not very keen on a third observatory in our small country. The reaction of the disappointed young astronomer was bitter: 'As a result of your advice a decent instrument will be refused to Groningen for an indefinite time. Throughout the Netherlands, the astronomer in Groningen will almost be the only one who does not have the necessary instruments for scientific research.

4.4 Marriage

Kapteyn initially lived in Groningen at the address Peperstraat 116 (see number 1 in Fig. 4.7); he worked in the Academy Building (A). His sister Cornelia Louise Alexandra ('Cor') (1855–1909), four years younger than Kapteyn, lived with him most of that time. She took care of the household for him, but also took lessons in German and English language and literature. We know quite a lot about this episode through the letters Kapteyn wrote to his future wife Elise, most of which was published (untranslated) in *Lieve Lize: De minnebrieven van de Groningse astronoom J.C. Kapteyn aan Elise Kalshoven, 1878–1879* by Klaas van Berkel and Annelies Noordhof-Hoorn (see the Preface for details). Maybe historians have less of a problem with that, but my experience, when reading these intimate and private letters, which were never intended for other people's eyes, is that it feels as an invasion of their privacy. I will therefore not quote

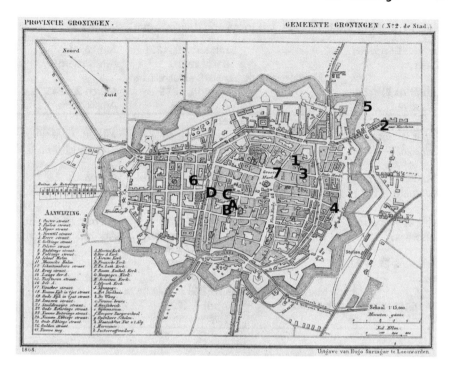

Fig. 4.7 Map of Groningen from 1868 with the locations where Kapteyn lived (figures) and worked (letters). 1 = Peperstraat 116 [now 11] (1878–1879), 2 = Winschoterkade WZ, X 132 (1879–1885), 3 = Oosterstraat OZ, D 42 (1885– < 1891), 4 = Heerestraat S 6a, later 113a (< 1891–1906), 5 = Emskanaal/Oosterhaven ZZ 16a (1906–1910), 6 = Ossenmarkt 6a (1910–1920), 7 = Grote Markt 35 (Hotel De Doelen) (1920–1921), A = Academy Building, B = Physiological Laboratory (1886–1896) and Astronomical Laboratory (from 1911 onward), C = Oude Boteringestraat (1896–1904), D = Oude Boteringestraat/Spilsluizen (1904–1911). From Groningen University Library and Municipal Atlas Jacob Kuyper (1865–1870) [46]

extensive parts of it (Table 4.1). For genealogical details of Jacobus Kapteyn, Elise Kalshoven and their parents, see Table (4.1).

It appears from these 'Love Letters' that Jacobus Kapteyn and Elise Kalshoven had no relationship at the time Kapteyn started as a professor in Groningen. But on August 13, 1878, Kapteyn made a marriage proposal to her by letter from Utrecht. Apparently he had spent part of the summer in Utrecht and their relationship had deepened in that time. Kapteyn insisted on August 16 as the date of the start of their relationship, so apparently Elise Kalshoven wrote back and accepted the proposal on that date, i.e. almost immediately after receipt of the letter. Mrs. Kapteyn maintained the starting date of the courtship as August 13, because the wrapper around the bundle in which the letters were

Table 4.1 Jacobus Kapteyn, Elise Kalshoven and their parents

Jacobus Cornelius Kapteyn	January 19, 1851 Barneveld	Jun 18, 1922 Amsterdam
	Married on July 17, 1879 at Utrecht to	
Catharina Elisabeth Kalshoven	June 19, 1855 Amsterdam	March 2, 1945 Amsterdam
	Parents	
Gerrit Jacobus Kapteyn	January 4, 1812 Bodegraven	July 26, 1879 Barneveld
	Married on March 12, 1837 at Bodegraven to	
Elisabeth Cornelia Koomans	November 5, 1814 Rotterdam	November 24, 1896 Barneveld
Jacobus Wilhelmus Kalshoven	June 18, 1813 Nieuwer-Amstel	February 14, 1869 Abcoude-Proostdij
	Married (I) on June 12, 1838 at Amsterdam to	
Catharina Elisabeth Brandt	June 30, 1814 unknown	November 9, 1849 Amsterdam
	Married (II) on June 1, 1853 at Zwolle to	
Henriëtte Mariëtte Augustine Albertine Frieseman	April 7, 1822 Harderwijk	April 13, 1895 Dieren

kept she had written on the outside: 'Blessed memory, read through on August 13, 1933–1934–1935–55 years after the day of engagement'.

A number of themes can be identified in this correspondence. The first is Kapteyn's problems in his new position. He complained extensively about the failure of obtaining an observatory. In November 1878 he spoke with concern about the rumors that the government wanted to close down one of the universities, and the presumption that this would be Groningen. Perhaps he would be transferred to Leiden and would have to work under Hendricus van de Sande Bakhuyzen again. He is also very nervous before giving lectures; he had extensively practiced his inaugural speech according to Henriette Hertzsprung–Kapteyn's biography [1]. 'Up to 30 times he recited his speech to two sisters, who patiently represented his audience and made comments from the two most distant corners of the room.' When fellow professor Mees was ill for a longer period of time he had to replace him at exams and PhD defenses. He felt rather insecure about that and it also took up a lot of his time. There was one Onnes, in whose thesis research he had played some role, in an area of which he felt he knew 'very little'. This was of course later Nobel laureate Heike Kamerlingh Onnes (1853–1926), who a few days before Kapteyn's marriage defended a thesis on new evidence for the rotation of the Earth.[1] So this was

---

[1] Mees had been the thesis supervisor, but must have been unable to fulfill that role.

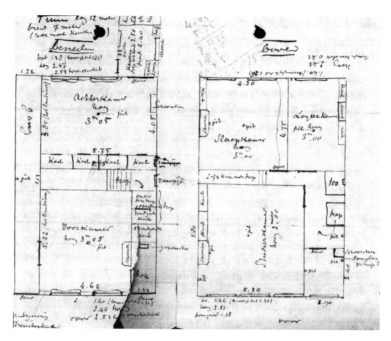

Fig. 4.8 The layout of their home, as Kapteyn, who imagined it and sent it to his fiancée Elise Kalshoven with a letter to her dated January 24, 1879. Reprinted from the *Love Letters*

not very far from his own field of research; his complaint that he knew too little about it seems to indicate an unjustified lack of self-confidence.

Kapteyn also asked his fiancée to write in English. Apparently he was less familiar with that language. At the boarding school in Barneveld French was spoken at the dinner table and German was in that period the lingua franca of science; until after the turn of the century Kapteyn himself published mainly in German. He even described Elise Kalshoven as half-English. From the letter of February 23, 1879: 'I was actually foolish enough to start an English epistle. What a foolishness. What are people supposed to think of me: English housekeeping, English clothes and curtains, English sisters, an half-English wife and now English letters as the crown to my work?' His sister Albertina Maria ('Bertamie') lived in England and their intended furnishing of their house was apparently English. Later, Kapteyn was one of the first scientists to show an Anglo-Saxon orientation and publish in English, which was still unusual in the Netherlands at the beginning of the twentieth century.

Another theme is the search for housing for after they would marry. The letters contain an extensive inventory of the number of rooms they would need (see Fig. 4.8). He starts with ten, but brings it back to five or six (two guestrooms indeed seems a bit much). But there must be a garden. Kapteyn also notes in his letter: 'I have not yet found the constellation M.G. because it has not been clear yet. Tonight I will look for it again.' I think the handwritten letter has 'M.Q.' (rather than M.G.), indicating the constellation 'Muur Quadrant' (wall quadrant) that is no longer in use. Indeed in October (the date of the letter) it would be in the evening above the horizon from Groningen.

Eventually he found and rented a house that was still under construction, on the Winschoterkade (2 in Fig. 4.7), away from the center of the town, where professors normally lived. That location would have a lower rent, although it was still considerable: 500 guilders per year for both floors (comparable to about 11,500 € purchasing power now), on an annual salary of 4,000 guilders. The house no longer exists; the address would now have been Winschoterdiep, along the canal that runs out of the city to the southeast. Elise Kalshoven visited Groningen (Kapteyn had to spend the night elsewhere, because at that time fiancés were not allowed to sleep under one roof) and they made all the preparations for their wedding. Figures 4.9 and 4.10 show Kapteyn and his fiancée not long before their marriage; this picture of Kapteyn is the only one I know of where he does not wear a mustache.

And that brings us to the third theme: the permission of Kapteyn's parents for the marriage. At that time, such permission was required until one had reached the age of 30. Jacobus Kapteyn and Elise Kalshoven had intended to marry only in a civil ceremony and would not marry in church; this was reason for the parents Kapteyn to announce that they would refuse their permission. This important issue takes up a large proportion of the letters. The parents Kapteyn felt that their presence at the marriage would implicitly endorse their son's principles. Kapteyn refused to renounce these principles, and found that his parents could simply make their disapproval known to everyone. He suggested that they could give their consent and stay away from the actual wedding ceremony. The mediation of brothers and sisters, who (just like the sister 'Cor' living with Kapteyn) partly chose the side of the parents, was of no avail, nor a personal visit by Elise Kalshoven to the parents Kapteyn in Barneveld.

Eventually the Kapteyns did marry in church. Father Kapteyn died ten days after the wedding, so maybe his health played a role in this decision. The letters mention the name 'Trenité'.[2] This was the priest of the Église Wallonne in Utrecht, Jean Gédéon Lambertus Nolst Trenité. This Calvinist church, which was visited by people from the southern Netherlands and France, and in which French was the official language, had at the time no more than about ten thousand members in the Netherlands. Perhaps the choice of this church was also part of some kind of compromise.

---

[2]The first acute accent on the surname 'Trénité' as spelled in Kapteyn's letters is incorrect.

Fig. 4.9 Jacobus Cornelius Kapteyn before his marriage to Catharina Elisabeth Kalshoven. This picture comes from Henriette Hertzsprung–Kapteyn's biography [1]

The marriage took place on July 16, 1879. The marriage certificate shows the signatures of the parents Kapteyn and the mother of Elise Kalshoven. The witnesses were: 'Johan Christaan Kalshoven, freight forwarding agent, thirty-five years of age, living in Rotterdam, brother of the bride, Simon Brouwer, notary, forty-six years of age, living in Maarssen, brother-in-law of the bride, Henrik van den Broeke, cashier, forty years of age, living here and William Burck, teacher at a Hoogere Burgerschool [HBS], thirty-two years of age, living in Apeldoorn.' Simon Brouwer was the person who originally brought Jacobus

Fig. 4.10 Catharina Elisabeth Kalshoven before her marriage to Jacobus Cornelius Kapteyn. This picture comes from Henriette Hertzsprung–Kapteyn's biography [1]

Kapteyn and the Kalshoven family together and William Burck was one of Kapteyn's bests friends from Leiden. The name van den Broeke has not been encountered before. He turns out to be the husband of Elisabeth Henriette Andreé Wiltens (1846–1940), the elder half-sister of Kapteyn's great friend from Utrecht, Willy Andreé Wiltens. Kapteyn probably had wanted to ask his old friend to be a witness, but he was living at that time in Padang, Sumatra, where he had bought a law firm. Apparently he could not afford the cost of the voyage to the Netherlands for the occasion, or did not have the time, and had arranged that his brother-in-law would take his place for him.

5

Nearly Half a Million Stars

Stars everywhere.
So many stars that I could not for the life of me understand
how the sky could contain them all, yet be so black.
Peter Watts (b. 1958).

His path was marked by the stars in the southern hemisphere.
Paul Simon (b. 1941)

By the end of 1879 Kapteyn had settled down. By now he had been a professor for almost two years, was married and together with his wife had set up a household. So, what now? There still was no observatory, and it did not look as if there was ever going to be one. It is sometimes been stated that he spent the following years in relative idleness, until at the end of 1885 he heard by chance of a proposal to map the southern sky with photographic plates and in an impulse offered his help. This suggestion, as well as the idea that he had accidentally ended up in astronomy, is incorrectly presented in Henriette Hertzsprung–Kapteyn's biography [1] and in a number of obituaries. It is true that he did some work that was not directly related to astronomy, but he certainly also did important astronomical research. I will start with the first.

Peter Watts is a Canadian science fiction writer and biologist.
From the song *African skies* of the album *Graceland* by Paul Simon.

P. C. van der Kruit, *Pioneer of Galactic Astronomy: A Biography of Jacobus C. Kapteyn*,
Springer Biographies, https://doi.org/10.1007/978-3-030-55423-1_5

5.1   Tree Rings and Mathematical Series

In his early years, Kapteyn researched systematic long-term patterns in astronomy and in the weather. Thanks to his teacher Buys Ballot, he was probably quite familiar with meteorological issues. Long-term weather was barely studied and long series of observations were not or hardly available. The weather forecast was in its infancy. Kapteyn's interest focused on whether there were any changes to be found that would recur periodically on timescales of a few years. That would bring the understanding of long-term weather patterns, and the facts that played a role in it, within reach. The underlying question was whether, as with solar and lunar eclipses, there was a regularity or pattern in the occurrence of weather.

Eclipses are based on the systematics of the orbits of the Earth and the Moon. The orbit of the Moon in the sky is not the same as the ecliptic (that of the Sun as seen from Earth in the course of a year). This is because the orbit of the Moon makes an angle (of over five degrees) with the orbit of the Earth around the Sun. On the sky, the corresponding orbits of the Sun and Moon cross each other in two points called nodes. The Moon returns to the same node with a period of 27.21 days (this is called a draconitic month). Because of the changing relative position of the Moon with respect to the Sun, it assumes phases from New Moon, via First Quarter to Full Moon—when they are opposite on the sky—and Last Quarter. In this too there is a well-defined period, which is 29.53 days (the so-called synodical month). A solar eclipse occurs when the Moon passes a node during New Moon, and a lunar eclipse when it does so during Full Moon. The Earth, Moon and Sun then are on one line in space. Now 223 synodical months is almost exactly equal to 242 draconitic months, and therefore solar and lunar eclipses occur with a regular cyclical pattern, which will be maintained for centuries. It fails in the end because of the 'almost exactly' in the last sentence. The corresponding period is 6585 days and 8 h.[1] This is called the Saros Period, which was already known in ancient times. Because of that 8 h (one third of a 24 h day) the eclipse will occur at the same place on Earth after three Saros periods.

The idea of Kapteyn now was that there might be such patterns to be found in the weather as well, which could then be discovered for example in the thickness of year rings in tree trunks. To investigate this, he collected slices of trees from various regions, such as the Trier area in Germany. However, he found the results too uncertain to publish. It was not until much later, in December 1908, that he gave a lecture on the subject in Pasadena, California, and published an article in the local newspaper, the Pasadena Star. Some time later, a slightly

---

[1] A period of 6585 days is equal to either 14 ordinary and 4 leap years plus 11 days, or 13 ordinary and 5 leap years plus 10 days.

more extended version of the newspaper article appeared in the Netherlands in the form of a separate publication. He had succeeded in collecting an almost continuous series of measurements of the thickness of annual rings between 1770 and 1880 for the trees in the Trier area. Those thicknesses correlated very well with meteorological data as far as available, such as the number of days of rain in each year or the total amount of precipitation. He also found a clear periodicity of 12.4 years and noted that this was significantly different from the solar cycle of about 11 years which is related to the number of sunspots. He gave his lecture in California because he wanted to point out that with the use of sequoia trees one could collect this kind of data over a thousand or more years. Incidentally, the period of 12.4 years has never been confirmed, as far as I am aware.

Kapteyn also worked with his brother Willem (Figs. 5.1) in Utrecht on a mathematical problem. This concerned the properties of certain mathematical

Fig. 5.1 Willem Kapteyn, brother of Jacobus Kapteyn, was a professor of mathematics at Utrecht University. This picture comes from the *Album Amicorum*, presented to H.G. van de Sande Bakhuyzen on the occasion of his retirement as professor of astronomy and director of the Sterrewacht Leiden in 1908 [47]

series, the 'higher order sines'. You can represent any mathematical function (such as the sine of an angle or the logarithm of a number) as a sum of regular terms that become smaller and smaller, so that you can calculate a reasonable approximation with the first ones. This series expansion has purely mathematical significance. The Kapteyn brothers published two long articles in French about it, with dozens of pages full of formulas. Properties of mathematical series was a specialty of Willem Kapteyn, who became an authority in it. He published a lot about it during his career, mostly in French. So William must have done most of the work. For those interested in mathematics, I briefly summarize what higher order sines and cosines are in Appendix A.3.

5.2   Kepler's Equation

However, in his first years in Groningen, Kapteyn also did original and significant astronomical research. I start here with his work on Kepler's equation, which he published in 1883. For that publication he chose a fairly new magazine, *Copernicus*, which was published in Ireland. It did not exist for very long; after less than three years it was terminated. It presented itself as an international journal and published papers in French and German in addition to English. Kapteyn was used to German as being the language in which astronomical literature was published, at least on the European continent, where the German Astronomische Gesellschaft set the tone; many publications went to the *Astronomische Nachrichten* published by this Gesellschaft. Kapteyn's paper, and the next one on absolute declinations, both were in German.

Kepler described around 1600 how the orbits of all objects in the Solar System (except for the disturbances by the planets among themselves of course) could be described. These are elliptical, with the Sun in one of the foci and the planets move faster when closer to the Sun; there also is a relationship between the size of the orbit (or mean distance from the Sun) and the period in it. Even before the end of the seventeenth century Newton showed that Kepler's laws were a direct consequence of his theory for gravitation. If one knew the characteristics of the orbit and at a certain moment the position of the planet, asteroid or comet in it, it was possible using Kepler's laws to calculate for any given moment where that object would be in its orbit. However, there was a serious problem. One had to solve a certain equation (named after Kepler), which was very labor-intensive. Methods had been devised, but it remained still very time-consuming. In Appendix A.4 I give a bit of an explanation for readers who want to know a little more about this (see the online version of the appendix for a bit more algebra). It was a serious problem because observatories simply did not have the manpower to do the necessary calculations. There was a real danger that asteroids were discovered but lost again before the orbit had been determined sufficiently accurately. It was not until the 1890s that the Astronomisches Rechen-Institut in Berlin,[2] under the directorate of Friedrich Tietjes (1834–1895) and Julius Bauschinger (1860–1934), assumed responsibility for this; from then on, these calculations have been made there systematically and the results kept up to date. Until the time of electronic computers, it remained a labor-intensive job to determine the orbits of comets and asteroids.

To Kapteyn, with his mathematical background, this looked like a good challenge. His method used mathematical series, as he studied with his brother Willem. However, these were other series than the ones they had published on and, as far as I know, Willem did not contribute to this. Kapteyn considered his work a real improvement, but closer inspection shows that it was marginal at best. It is ironic that later a mathematical series was discovered, named after Kapteyn (but then Willem), that eventually would turn out to facilitate a faster algorithm of solving Kepler's equation.

5.3   Positions and Parallaxes of Stars

Kapteyn's most significant work in the early 1880s concerned his most fundamental interest: the spatial distribution of the stars, the 'Construction of the Heavens'. In order to be able to study this, one needs catalogs and for this precise measurement of the positions of stars to great accuracy is required. We saw above that the position of a star in the sky in a catalog is defined by two numbers. The first is the right ascension, which is determined from a precise time measurement of the passage of a star through the meridian (the moment when the star is exactly in the south). The second, which is perpendicular to it, is the declination, which is determined by the angle above the horizon of the direction of the star at that moment. However, the measurements of declinations posed a major problem for an observer.

That problem is twofold. The first point concerns the bending of the telescope's tube. A telescope is not perfectly stiff; it will bend under its own weight, so the measurement of the angle of a star above the horizon will be too large. That would not be too bad if you could correct for it, but it was not easy to determine how much that deflection for any particular telescope is. Secondly, there is the refraction of light. As soon as light from a star enters the atmosphere, the path of the light is refracted, and that refraction increases when a

---

[2]This institute has been in Heidelberg since 1945.

star is seen closer to the horizon. The starlight will follow a curved trajectory and the star will also appear to be higher above the horizon than it is in reality. The existence of this refraction was known, but no one knew how to correct for it. It must depend on factors as barometric pressure, temperature and humidity. Both these effects are actually absent in the zenith and increase as you get closer to the horizon. The result of all this was that different star catalogs were quite similar to each other in terms of right ascension, but sometimes differed considerably in declination. That was an urgent problem that Kapteyn wanted to address. He called it the issue of absolute declinations.

His clever solution consisted of two parts. First, he assumed that for an observatory the polar altitude was known exactly. The trick then was to determine the absolute declinations of a network of stars across the sky, from which the declinations of other stars could be determined by a *relative* measurement over a small angle. Kapteyn now devised a clever technique that was independent of telescope bending and refraction. This consisted of measurements of the 'azimuth', which is the angle that the projection of a star from the zenith on the horizon makes with the south (or the north), and timing of the passing of a star through the prime vertical (comparable to the meridian, but now from east to zenith to west on the horizon). Furthermore, he also made use of two stars that pass the meridian at about the same height above the horizon, one in the south and the other in the north and subtracted their altitudes above the horizon. In these measurements the effects of the refraction and telescope bending are then about the same and cancel when you subtract the two measurements from each other. Kapteyn's complicated technique did not work for every declination, but there were several declination zones in the sky where it did work, and that was enough to determine a network of stars with absolute declinations. In his article on this subject, again published in the *Copernicus* magazine, Kapteyn presented in detail the mathematical techniques that could be used to set up such a network from such measurements. This had to be done of course for the same set of stars around the equator from a northern and a southern observatory to tie together declinations in both hemispheres.

The problem was not yet solved but reduced to a determination of the absolute polar height (altitude) at the observatory where the measurements were done. That could be done, as Kapteyn showed, using a set of three stars. However, these had to meet special conditions. A first star had to be circumpolar (close to the pole so that it never sets) and the other two had to have declinations such that they passed the meridian at roughly the same height above the southern horizon when the first star in the north did so above and below the pole. These then had to have right ascensions, such that these passages took place at about the same time. In this method the telescope bending and refraction do not play a role, because again differences in altitude were used and then these effects cancel in the subtraction. Kapteyn showed that with his technique is was possible to determine the absolute altitude of the pole above the horizon. These conditions were so specific that in practice no such combination of three stars was available, but Kapteyn showed that one could still get good results when observations of a large number of similar approximate combinations of stars, scattered all over the sky, were used.

Kapteyn did not stop with merely presenting a method. In the summer vacation of 1882 he was allowed to test his method with an instrument in Leiden (because of bad weather he partly used measurements he had taken for other purposes in 1875). He used a so-called universal instrument (see Fig. 5.2), which was actually designed for use in geodesy and surveying. It became, however, used extensively for teaching astronomy. With his observations he showed that the method did not only work in theory, but could actually be implemented in practice. It was a complicated exercise, but it would be able

Fig. 5.2 The Universal Instrument of the the Sterrewacht Leiden, which Kapteyn used to test his method to determine the absolute polar height. It was produced by J.G. Repsold and Sons, in Hamburg, and procured in 1853. The dimensions are 50 cm by 78 cm. It is now in Museum Boerhaave in Leiden [48]

to solve the problem of absolute declinations. So this was not only rather fundamental, but also very inventive and clever work by Kapteyn.

What was most urgent for Kapteyn's aim of studying of the arrangement of the stars in space was measurements of *distances* of stars, and that was the next problem that Kapteyn addressed. The only method available until then had been measuring the projection of the motion of the Earth around the Sun in the position of a star. This manifests itself as a small ellipse in the sky and can be measured for a nearby star relative to some, generally faint, stars in the background which would be at larger distances. I have already discussed this above; there we also saw that the parallax of the nearest star α Centauri is only $0\rlap{.}''75$, so less than a second of arc.

Because parallaxes are so small, they were known in the 1880s only for a very small number of stars. Of course Kapteyn needed many more parallax measurements to be able to carry out his studies of the spatial distribution of stars. We already saw in proposition XV of his dissertation, which dealt with the relation between the distance of a star and its proper motion in the sky, that this issue has preoccupied him from an early age onward. Now proper motions result from the velocity of a star through space; the proper motion will generally be larger when a star is near to us. But it can measured very accurately if one is prepared to wait longer, because the displacement on the sky increases with time. Because star positions were available from antiquity the time available covered several thousand years, Edmond Halley discovered proper motions in 1718.

Because the orbital plane of the Earth makes an angle of only about $23°$ with the equator, the effect of parallax occurs mainly in the direction of the right ascension of a star; so it should in principle be possible to measure it using the timing of the meridian passage of a star. Of course you do this then relative to a number of fainter stars in the sky that in general would be further away. Now others had thought about this before, but nobody had ever managed to perform this measurement. That is because the effect is so small. For example, for a star with a parallax of $0\rlap{.}''1$, which is only 7.5 times further away than α Centauri, the effect is of the order of a few hundredths of a second if time, so you must be able to determine the meridian passage with at least that accuracy! Kapteyn was not deterred by this.

Hendricus van de Sande Bakhuyzen gave Kapteyn permission to use the meridian circle of the the Sterrewacht Leiden (see Fig. 3.6) during several academic vacations in 1884 and 1885. The measurements had to be done six months apart, and then repeated again after another six months to correct for the proper motion. Kapteyn selected fifteen stars, of which he suspected on the basis of their proper motions that the distances were relatively small. But how

Fig. 5.3 The Registrir-Apparat built by Mayer and Wolf, a specimen of which was available at the the Sterrewacht Leiden and which was used by Kapteyn for his measurements of parallaxes of stars [49]

can one measure times accurate to a fraction of a second? He used a so-called 'Registrir-Apparat', through which a paper strip moved, in which the device automatically punched a hole every second (see Fig. 5.3). The observer could also do this by pressing a button when the observed star in the telescope crossed a cross-wires in the field of view. The device had a mechanism that ensured that the little pen would not immediately pull a tear in the paper if one pressed the button too long. Even then, one could only achieve the required level of accuracy by doing with extreme consistency a number of measurements and then averaging them.

Kapteyn indeed managed to measure some parallaxes. With a single measurement he succeeded in obtaining an accuracy of between one and two tenths of a second of arc, with repeated measurements even 0″.03. Comparison of his results with modern values yields the astonishing conclusion, that his parallaxes are as accurate as he claimed (for details see my scientific biography of Kapteyn [5]). He must have been a very accurate and careful observer. But the conclusion was disappointing; the distances of stars were so large, that determining parallaxes in large numbers for a statistical study of their positions in space was for the time being impossible.

5.4   Collaborating with David Gill

Sir David Gill (1843–1914) (Fig. 5.4) was born as the son of a clock-maker and grew up in Aberdeen. He also studied there, among others, with the famous physicist James Clerk Maxwell (1831–1879), and then worked for some time in his father's company. But he got tired of the trade and started doing astronomical work, first of all on the private observatory of James Ludovic Lindsay, 26th Earl of Crawford and 9th Earl of Balcarres (1847–1913). Lindsay was president of the Royal Astronomical Society. David Gill's specialty was the maintenance and construction of instrumentation; it was because of these skills that he took part in the expedition to Mauritius in 1874 to observe the transit of Venus in front of the Sun. In 1877, he took part in an expedition to Ascension, an island in the South Atlantic Ocean, to make observations of Mars when it was very close to Earth. Both expeditions were set up to better determine the

Fig. 5.4 David Gill (1843–1914). This photo comes from Henriette Hertzsprung–Kapteyn's biography [1]

scale of the Solar System. In 1879 he was appointed 'Her Majesty's Astronomer' at the Royal Observatory at the Cape of Good Hope, where he would remain until his retirement in 1906.

It has been suggested, in the biography of Kapteyn's daughter as well as obituaries and other biographical articles about Kapteyn, that the collaboration between Kapteyn and Gill arose when Kapteyn read an article during the Christmas holidays of December 1885, in Leiden, in which Gill described his plans to construct a star catalog of the southern sky using photographic exposures, and sought help with this. Kapteyn would have spontaneously offered his services. But that is not at all how it happened. Such an article by David Gill does not even exist; it would have been published in the British journal *the Observatory*, but there is no such an article in it. What happened is the following. Kapteyn himself sent his article on the measurement of absolute declinations and polar height to Gill with an accompanying letter dated April 30, 1884. In it he refers to an article that does exist, by David Gill in the *Vierteljahresschrift* of the Astronomische Gesellschaft on the question of absolute declinations. Gill wrote back (with some delay) that he found the article 'a model of clarity'. He also suggested setting up an observing program to put Kapteyn's ideas into practice.

Interestingly, David Gill wrote some time later that he had Kapteyn's article translated from German into English because he was not very good at languages; he added that 'it is all clear to me <u>now</u>' (!!). The annual reports of the Cape Observatory do mention observations for the Kapteyn project for several years, but as far as I know no final results were ever published. In 1882 Gill had made photographs of the 'large comet' that was visible that year, and he had noticed that there were large numbers of stars visible on the plate (Fig. 5.5). Now above we saw how the *Bonner Durchmusterung* had been produced in the northern hemisphere, which had yielded relatively accurate positions and brightnesses of stars up to at least magnitude 9–10. This had then been extended from Bonn to all stars that were visible from there in the *Südliche Durchmusterung*, but most of the southern sky had not been mapped to any reasonable accuracy for all but the brightest stars. It was planned to do this from the Córdoba Observatorio Astronómico in Argentina, but it had not gotten very far yet. And the intention was that, like the northern duchmusterungs, it would be done star by star with a meridian circle, which could take many years. David Gill realized that in principle it could be done much faster with photographic images. He therefore started taking the first exposures with a specially built telescope with a lens of 15 cm and a focal length of 137 cm (see Fig. 5.6). But he had absolutely not the time nor the manpower to measure up those plates.

1882 Nov 7ᵈ.

Fig. 5.5 The great comet of 1882, photographed by David Gill, on which he noted the presence of many stars which could be used for making a southern star catalog. From the website of the South African Astronomical Observatory [50]

He did not directly ask Kapteyn to take on this task, but in their correspondence he regularly returned to the project. Gill asked Kapteyn for advice on how to measure the plates and to estimate with what accuracy that would be possible; and he regularly complained about how little time he had for all sorts of things. It was not until December 1885 that Kapteyn took the bait and offered that he would measure the plates. He estimated it would take him seven years. He asked the directors of the Leiden and Utrecht observatories, Hendricus van de Sande Bakhuyzen and Oudemans, for advice, but although they strongly advised him against it, Kapteyn decided to go along with David Gill. Thus began the project that was to be called the *Cape Photographic Durchmusterung* or *CPD*.

Kapteyn managed to raise some funds, although his applications were also regularly rejected, among others by the well-known Teylers Foundation. But in the end he received support from both the government and organizations

Fig. 5.6 The instrument at the Cape Observatory that was used for taking the plates for the *Cape Photographic Durchmusterung*. Plate I in the *CPD*, Kapteyn Astronomical Institute, University of Groningen.)

such as the Batavian Society of Experimental Philosophy in Rotterdam and the Province of Utrecht Society of Arts and Sciences. He had no room to put up a measuring apparatus, but for this he was offered the use of two rooms in the (Groningen) Physiological Laboratory of his friend Dirk Huizinga (1840–1903). An experienced assistant was too expensive, but he found a solution for that too by appointing a young man who had just come from the school for crafts and technical workmen, and by training him himself. That was Teunis Willem de Vries (1862–1937), who would continue to work for Kapteyn and his successor Pieter van Rhijn until his retirement.

Kapteyn devised a very clever method of measuring the plates. There was of course no real experience with measurements of star positions from photographic plates. The obvious method was to measure the position of each star on the plate as the vertical and horizontal distance from one of the corners, or maybe distance and direction from the center of the plate. This then should include stars which had a known position on the sky to calibrate the astronomical coordinates. Gill had already asked Kapteyn in one of his letters, even before Kapteyn had agreed to cooperate, to weigh these two methods against each other. Kapteyn had chosen the first one at that time. Actually, in neither of these two cases it was straightforward to convert those measurements on the plate into coordinates on the sky. That required a lot of calculation, although it gave the most accurate result. In fact, it was more accurate than was necessary for a catalog similar to the *Bonner Durchmusterung*. The method that Kapteyn devised in the end, and which he called the 'parallactic method', was perhaps somewhat less accurate, but certainly acceptable for the *CPD*. And a lot faster and much more straightforward. Kapteyn used a small telescope (which had previously served other purposes) to look at the plate, which was mounted in a plate holder (see Fig. 5.7). The arrangement consisted entirely of older and

Fig. 5.7 Kapteyn's measuring device to measure the plates for the *Cape Photographic Durchmusterung*. On the left the plate holder, on the right the measuring telescope. The distance is not to scale. University Museum Groningen

often discarded parts, partly from the workshops of other departments of the university.

Kapteyn's clever parallactic method was the following. If you now make the distance between the plate and the measuring telescope exactly equal to the focal length in the telescope that was used for the exposure of the plates (137 cm), then the orientation of the telescope on its two axes would directly correspond to celestial coordinates. The difference with the readings for the center of the plate were then directly the difference in right ascension and declination between the star and that center. The set-up can be seen in Fig. 5.7; the photographic plate was orientated in such a way that the north in that figure was to the right. Note that the distance between the plate and the telescope is not to scale. In working order, the plate holder contained two separate plates of the same part of the sky one behind the other, but slightly shifted. In this way plate faults and other iniquities could be identified immediately. Also, from the size of the dark spot that a star left behind in the photographic emulsion, the brightness of that star was estimated.

David Gill had started exposing plates in 1885 and completed the full set in 1888, although he still took more plates at Kapteyn's request (see Fig. 5.8 for a box with a few of the actual plates). The measuring in Groningen began on October 28, 1886 with an exposure centered on the South Pole; the last measurement was on June 11, 1892. There were always two people involved, often other assistants who Kapteyn could pay from various sources. (Apparently the salary was not that good, because there was quite a bit of turnover.) In the years that followed, extra measurements were taken sometimes of fields that Gill

Fig. 5.8 Box with plates for the *Cape Photographic Durchmusterung*, which are still kept in the University Museum Groningen

had exposed again because the first ones were of insufficient quality. Initially Kapteyn himself was involved in the execution of the measuring, but later he left it to de Vries and his assistants. Numerical work still had to be done before a catalog could be produced. Continuous quality control was also necessary to keep the whole consistent. The story goes that inmates from the local prison helped to measure the plates, but that is unlikely (they mostly had very little schooling) and no record of it has been found in the prison's archives.

The *Cape Photographic Durchmusterung* was finally published in three thick volumes in 1896, 1897 and 1900, as parts of the *Annals of the Cape Observatory*. In total it contained the positions and magnitudes of 454,875, almost half a million, stars. The completeness was up to magnitude 9.2, positions were accurate to 3 or 4 arcseconds and magnitudes up to 0.06 mag (corresponding to 6% in brightness). The number of plates measured eventually amounted to 613, but more than a thousand had been rejected (in most cases after they had already been measured); they were listed in the CPD in order for others to be able to search for variable stars later on.

This work received quite a lot of attention from astronomers worldwide. Kapteyn's star was already rising through his participation in international conferences and publications, but the CPD made him a leading astronomer. He was almost fifty in 1900, when the last volume came out, so it took some time before he got to that status. But he still had twenty years to go until his retirement. In a way, all the above had only been preparation for the real thing to come.

5.5   France and the International Organization of Science

The production of the *Cape Photographic Durchmusterung* was quite an undertaking. But an even larger, international project arose at about the same time, and eventually even became a threat to the completion of the CPD. That was the *Carte du Ciel*. But before I into that in more detail, I must first say something about the French role in the international organization of science.[3]

In the second half of the nineteenth century, France was very keen to structure global science and technology according to its insights. Some French initiatives were accepted internationally, but often things turned out differently than the French had imagined. Here are a few examples. In the course of the nineteenth century, increasing mobility, especially by train, made it in-

---

[3]An excellent account of this can be found in the book *Einstein's Clocks, Poincaré's Maps, Empires of Time* by Peter Galison [51].

creasingly necessary to coordinate times and clocks internationally. Until then, everyone used their own local time, related to the local geographical longitude, in which the clock was basically set in such a way that on average throughout the year the Sun was exactly in the south at noon. I digress briefly in order to explain why I used the wording 'on average'. It is not that simple in practice, because there is the so-called 'equation of time', which is the property of the Sun to be on the meridian too early or too late relative to the clock. This difference varies in the course of a year. Its cause is that the Earth in its orbit around the Sun does not have a constant speed and also because the rotation axis of the Earth is not perpendicular to that orbital plane. The difference can be up to 18 minutes, but of course averages out over a whole year. This can be seen in the *analemma*, the distorted figure eight that is often depicted on old maps and sundials. It shows up when we take a photograph of the Sun in the sky every day at exactly the same time *on the clock*; if this is done for a full year on a single plate, then the images of the Sun on it form this figure. The vertical, long axis of the analemma is created by the seasons, which makes the Sun high in the sky in the summer and low in the winter; the horizontal axis by the equation of time.

In order to achieve greater uniformity, the idea had been put forward to divide the Earth up into 24 zones of 15° in geographical longitude, whereby the time within each zone would be the same everywhere and would differ by an hour from the zones next to it. For practical reasons the zones were not chosen as exactly 15°, but national borders were used. This system had originally been proposed in 1879 by Sandford Fleming (1827–1915), a Scottish-born Canadian engineer, but was also suggested by others independent of him. The question was then where the meridian would run that would define longitude zero. In October 1884, after long and bitter discussions, the decision was made in Washington. The zero meridian ran through the Royal Observatory in Greenwich and not through l'Observatoire de Paris. This 'defeat' was difficult to accept for the leader of the French delegation, the astronomer Pierre Jules César Janssen (1824–1907).

The Netherlands (and France and most of western continental Europe) are in the the Central European Time zone, one hour later than Greenwich Mean Time (nowadays called Universal Time). For the Netherlands the current situation actually is less favorable; on average (except for the equation of time) the Sun is in Utrecht—for example, more or less the center of the Netherlands—on the meridian at 12:39:32 (and one hour later during Daylight Savings Time). Had the French been given the zero meridian, it would have been 11:49:53.

In the course of time, the French also proposed a complete decimal system, which was only put into use in a few respects. Shortly after the French Revolution a system was introduced in which months were divided into 'decades' of ten days instead of weeks of seven, and one day into 10 h, and so on. Pierre-Simon Laplace (1749–1827) had played a part in this proposal. Napoleon abolished it. In Washington, Janssen advocated a universal system of decimal division of angles and time. Later, this was also worked out by a committee led by the mathematician Henri Poincaré (1854–1912). For example, a day then would have 10 h, each consisting of 100 minutes and each minute of 100 seconds. The differences with the current time-line are not even dramatic. A hundredth of a day would then be 10 minutes, and would be equal to 14.4 current minutes, so almost a quarter of an hour; and a hundred thousandth of a day (the new second) would measure 0.86 current seconds. According to the proponents it was also more logical to divide a full circle into 400 units. A right angle is then 100 'degrees' and you divide that in a decimal system again. As we know, this never got accepted. However, France did succeed in introducing the meter via the system just mentioned, which was registered as one forty millionth of the Earth's circumference. But the second of time (and the hour and minute) was defined in the Système Internationale d'Unités (S.I.) in the old way. We all know how especially the United States still adhere to an archaic and complicated system of distances, weights and temperatures.

The only French success was that the standard meter and kilogram were stored in the Bureau International des Poids et Mesures in Paris, but these are hardly of scientific importance. In the meantime, the second and the meter have been defined on the basis of spectral transitions in atoms; the standard meter has only historical value. Things will soon be not much different with the kilogram; a redefinition in terms of atomic properties only awaits a somewhat more precise determination of the Avogadro's number, the number of atoms in a mole, a standard quantity of a substance. It is defined as the number of atoms in 12 grams of Carbon-12. Avogadro's number, named after Amedeo Avogadro (1776–1856) is about 6×10^{23}, so a 6 followed by 23 digits before the decimal point. To redefine the kilogram, one needs to know this number to at least the first 8 digits, one or two more than is currently known.

5.6   Carte du Ciel

David Gill had emphasized the potential of photography for astronomy on several occasions. The director of l'Observatoire de Paris, Amiral Ernest Amédée Barthélemy Mouchez (1821–1892), had been impressed by the presentations and writings of Gill, and the possibility of using photography to catalog stars. He shared this with the brothers Paul-Pierre (1848–1905) en Mathieu-Prosper

Henry (1849–1903), usually named Paul and Prosper Henry. The Henry's were excellent instrument and telescope builders and soon set to work to improve the technology for astronomical photography.

In view of the previous section, it will come as no surprise that France also saw a leading role for itself in this area. With the help of Gill and through him of Otto Wilhelm von Struve (1819–1905), director of the Pulkovo Observatorium near St Petersburg, Mouchez organized a large congress in Paris in April 1887 to forge plans for an international project to produce a 'map of the sky'. All of this under the direction and supervision of France. The objective was not only a catalog to fainter magnitudes than the *Bonner* and *Cape Photographic Durchmusterung*, but also an atlas (according to Gill at Kapteyn's suggestion!) printed on paper.

Kapteyn was present in Paris: see the group photo in Fig. 5.9 and some special individuals in Fig. 5.10. Among the prominent figures were Hendricus van de Sande Bakhuyzen and Jean Oudemans, who were allowed to sit in the front row. Kapteyn does not seem to be very comfortable from the group photo. There is another version of this photo on which Kapteyn and a few others are missing. Kapteyn's name does not appear on the table setting at the buffet, which was part of the congress.

That same month a group of gentlemen met to further set up the project. The *Carte du Ciel* was to be carried out by a large number of observatories in

Fig. 5.9 Group photo of the meeting planning the *Carte du Ciel* in 1887 in Paris. All attendants have put their signatures under the picture. Kapteyn is in the second row second from the left. From the archives of the Kapteyn Astronomical Institute, University of Groningen

Fig. 5.10 Some individuals relevant to this story as details taken from the group photo in Fig. 5.9. Top: on the left Kapteyn with above him the brothers Paul (left) and Prosper Henry; in the middle David Gill, on the right Anders Severin Donner of the Helsingfors Observatory near Helsinki. Below we see some prominent people, important enough to be allowed to occupy a chair: from left to right van de Sande Bakhuyzen, Oudemans and the big shots: William M. Christie, Astronomer Royal and director of the Royal Greenwich Observatory, Amiral Ernest Mouchez, host and director of l'Observatoire de Paris, Otto von Struve, director of the Pulkovo Observatorium near St Petersburg and Arthur von Auwers, secretary of the Berlin Academy. Kapteyn Astronomical Institute, University of Groningen

various countries. The United States did not participate, although there had been American representation at the congress in Paris. France dominated the project, which was to be coordinated from l'Observatoire de Paris. The minutes of the meeting of 1887 were published in full in French. A 'Permanent Committee' was set up under Mouchez. Studies were carried out on behalf of this committee and were also published in French, undoubtedly translated by the French if necessary. It was decided that the *Carte du Ciel* would consist of two parts. The first part, the *Astrographic Catalogue*, would contain the positions and magnitudes of all stars up to magnitude 11. The photographic plates would be exposed with as many special, identical telescopes as possible, under the same conditions, with the same emulsions and exposure times. The telescopes had an objective lens of 33 cm and were manufactured by the Henry brothers or by the Grubb company in Dublin, founded by Howard Grubb

(1844–1931). The emphasis on uniformity not only proved difficult to implement in practice, but also made the whole project unnecessarily rigid. Twenty observatories participated, none from the USA. According to Hendricus van de Sande Bakhuyzen, the Sterrewacht Leiden could not afford the purchase of the telescope and the materials for the execution, and did not contribute.

The production of the *Astrographic Catalogue* was not without problems. Some observatories were relatively fast, others took much longer to take the plates. The intention was that all exposures were taken three times on the same plate, each time with a slight shift of the direction in which the telescope pointed to the sky. In this way, stars could be distinguished from plate faults; and because the three observations had different exposure times, effects of overexposure were also under control. But all of this was a lot of work and measuring up the plates was extremely time consuming. The quality requirements were very high, and rightly so. The project comprised 22,000 plates; the catalog was only completed in 1961. However, completing is a misleading word, because most published volumes only contained measurements in rectangular coordinates on the plates, and it still was a huge job to convert these into right ascensions and declinations. Still, this exercise turned out not a wasted effort. When the European Space Agency launched the astrometric satellite Hipparcos in 1989, which carried out accurate measurements of parallax and proper motions of over a hundred thousand stars until 1993,[4] the *Astrographic Catalogue* was digitized and reduced to proper coordinates and made available electronically and was used as a 'first-epoche' measurement for the proper motions. As a result, in many cases the time interval over which proper motions could be measured increased to almost a century.

The second part of the catalog was a 'real' *Carte du Ciel*, i.e. a map of the sky in the form of photographic images. In the end, the idea was to produce these as copper engravings, but that soon turned out to be far too expensive. This part never got very far. In addition, a similar project was carried out around 1950, at least for the northern hemisphere, in the form of the National Geographic Society—Palomar Observatory Sky Survey. This was produced with a special telescope, the 48 in. (1.2-m) Schmidt telescope at Palomar Observatory in California. It was specially designed to take photographs of large pieces of sky. The *Sky Survey* was produced in pieces of sky measuring $6° \times 6°$ (with the *Carte du Ciel* it would have been $2° \times 2°$, so almost ten times smaller fields) on photographic plates of 14×14 inches (35.5 cm). The exposures date for the greater part from between 1956 and 1958 and were then printed on paper. Every astronomical institute still has a copy of these; the *Sky Survey* was very affordable. The southern sky was mapped in the 1970s with

---

[4] Later a catalog was published based on Hipparcos observations of 2.5 million stars, with greater accuracy.

similar telescopes from the Anglo-Australian Observatory in Australia, and the European Southern Observatory (ESO) in Chile. The results of these surveys are now publicly available electronically as the *Digital Sky Survey* [52] More recently this has been replaced by the *Sloan Digital Sky Survey* [53], which goes deeper but covers only a limited part of the sky.

Kapteyn actively participated in the *Carte du Ciel*, partly because he was (after some time) elected to the Permanent Committee. He tried to convince the leadership of the project to adopt his parallactic method for measuring the plates. The project actually did allocate funds to him to build a prototype to be used in the production of the *Carte du Ciel*, but eventually the method was rejected. The procedure of rectangular coordinates chosen by the Committee was indeed potentially more accurate, but if—as happened—the measurements are not reduced to celestial coordinates, it is not a very useful approach. Kapteyn also suggested carrying out the observational procedure in such a way that parallaxes and proper motions could be determined on a large scale. He proposed to do the exposures of each field three times on the same photographic plate, but contrary to the adopted procedure with six months intervals. In principle, you can then measure the parallax and proper motion from the displacement of the stars between those three times. But then you twice have to leave the plate half a year undeveloped and only develop it when the third exposure was done. This proposal of Kapteyn was not accepted either. Eventually Kapteyn stopped his activities for the *Carte du Ciel*. Figure 5.11 shows Kapteyn during the times when all this was happening.

The *Carte du Ciel* also became a threat to the *CPD*; this was mainly the result of some animosity between David Gill and William Mahoney Christie (1845–1922), Astronomer Royal and director of the Royal Greenwich Observatory (see Fig. 5.10). Christie was jealous of Gill and convinced the Royal Society to stop subsidies to Gill for the *CPD* in 1887. Among other things, he claimed that far fewer stars appeared on the plates than was expected. That was pure nonsense, as the final total number of almost half a million illustrates. When other countries offered to support Gill financially, Christie threatened to stop the English participation in the *Carte du Ciel* if these funds were accepted by Gill. In the end, Gill paid for the project out of his own pocket; his wife gave up her carriage and Gill even stopped smoking cigars! 'We will somehow manage it together even if I have to give up tobacco,—which God forbids', Gill wrote to Kapteyn. In the end, Christie's misconduct was exposed and he was deprived of his previous authority to advise on the Cape Observatory. Gill suffered financially, but Christie suffered in terms of reputation. By way of illustration: two important prizes in astronomy were (and are) the Gold Medal of the Royal Astronomical Society and the Catherine Wolfe Bruce Gold Medal

Fig. 5.11 Kapteyn, probably in his thirties. This illustration has been taken from Henriette Hertzsprung–Kapteyn's biography [1]

of the Astronomical Society of the Pacific. David Gill was knighted in 1900 and received the first of these medals in 1882 and in 1908 (to win it twice is extremely rare), and the other prize in 1900. Christie was also knighted in 1904, but was never awarded any of these very prestigious medals (nor any other important distinction).

5.7   The Kapteyn Household

Almost a year after Kapteyn and his wife were married, their first child, Jacoba Cornelia, was born; a year and a half later their second daughter, Henriette Mariette Augustine Albertine, followed. Another two years later their son Gerrit Jacobus was born (see Table 5.1). The registers at the municipality of Groningen also contain the statement of a stillborn child in August 1895. In the family circle the children were addressed with nicknames, respectively 'Dody', 'Hetty' and 'Rob'; those names were also used by family, friends and acquaintances. Since I am not in those categories, I will not use those nicknames here. Figure 5.11 shows Kapteyn around that time.

Jacobus and Elise Kapteyn had thoroughly studied the principles of child care and upbringing. Together they studied the book *The development of the child in body and soul: companion for mothers raising their first child* by the

Table 5.1 The children of Jacobus and Elise Kapteyn and their partners

Jacoba Cornelia ('Dody')	Groningen, May 31, 1880	Amsterdam, October 4, 1960
	Married July 14, 1906 at Vries to	
Willem Cornelis Noordenbos	Hallum, June 10, 1875	Amsterdam, August 18, 1954
Henriette Mariette Augustine Albertine ('Hetty')	Groningen, November 16, 1881	Utrecht, October 15, 1956
	Married (I) May 16, 1913 at Groningen to	
Ejnar Hertzsprung	Fredriksborg, October 8, 1873	Roskilde, October 21, 1967
	Divorced January 19, 1937	
	Married (II) April 17, 1937 at London to	
Joost Hudig	Rotterdam, March 18, 1880	??, June 4, 1967
Gerrit Jacobus ('Rob')	Groningen, December 14, 1883	Davos Platz, December 25, 1937
	Married January 30, 1918 at Groningen to	
Wilhelmina Henriette van Gorkom	Johannesburg, January 7, 1895	Utrecht, December 21, 1953

Amsterdam physician Gerardus Arnoldus Nicolaus Allebé (1810–1892), which was first published in 1845 and had become a standard work for young parents in the second half of the nineteenth century. In her biography of Kapteyn, Henriette Hertzsprung–Kapteyn [1] wrote:

> Together they had made a thorough study of the principles of child care. This was an exception in those days, but they were ahead of their time in their minds, and so they were able to put right the wise nosy midwife, who was used to undisputed rules according to old-fashioned recipe. This was no small thing indeed, for at that time the midwifes ruled fiercely and tolerated no other opinions. Kapteyn also opposed the doctor. He could follow the generally recognized authority only if his common sense said it was correct. When the child [the eldest daughter], who was one month old, had a severe bowel problem, and the doctor had prescribed more protein, he had the courage, and the strength, to reject this recipe. Logic told him that the weak stomach of a almost dying child would not tolerate proteins. He prescribed a diet of sugar water, limiting it to a minimum. The father and mother watched the result anxiously. What a responsibility they took upon themselves! But the child recovered and with that the confidence in his own wise insight grew.

But Kapteyn could also admit that he was wrong:

> Another of Kapteyn's theories was that a child naturally had no love for dolls, but that it was introduced to her by the parents. That is why he did not want his child to have a doll. So poor Dody had to go through life doll-less. Once, however, he saw that the child rocked an ugly dark Japanese doll that one could move up and down a stick, gently back and forth in her arms like a mother who was rocking her child. He did not say anything, but in the afternoon he came home from town with a package for Dody, in which the enraptured child found a beautiful doll.

After some years, probably in 1885, the family moved from the Winschoterkade to a more central location in Groningen, namely Oosterstraat OZ (east side) 42 (see Fig. 4.7, number 3). The Kapteyns probably occupied the upper two floors. Today, an interior design and furniture store is located on the ground floor. They did not live here for very long, because according to the archives of the municipality of Groningen they had moved to Heerestraat 6a in 1891, later renumbered to 113a (number 4 in Fig. 4.7). This building does not exist anymore. Apparently they also moved into an upstairs apartment there; they stayed there until 1906, when the children had gone to university.

Fig. 6.0 Kapteyn at age 35. This photograph has been taken from Henriette Hertzsprung-Kapteyn's biography [1]

6

Laboratory and Statistical Astronomy

'Messy', concluded Mr. Charles C. M. Carlier. 'And sloppy!'
It all was, if you considered it carefully, really quite messy, sloppy actually,
and unfinished –the creation and all that, the starry sky and all that,
even though it was the summer sky of his beloved Provence.
'Typical rush job', said Mr. Charles C. M. Carlier.
Hendricus Frederikus van der Kalle, alias Havank (1904–1964)

It has always irked me as improper that there are still so many people
for whom the sky is no more than a mass of random points of light.
I do not see why we should recognize a house, a tree, or a flower here below
and not, for example, the red Arcturus up there in the heavens as it hangs from its
constellation Boötes, like a basket hanging from a balloon.
Maurits Cornelis Escher (1898–1972)

The final part of the *Cape Photographic Durchmusterung* was published in 1900, the last year of the nineteenth century. Kapteyn, who was to celebrate his fiftieth birthday in January of the next year, had been working on it for more than a decade. But it would be wrong to think that he had not done any other work. In that year 1900, for example, he started the series *Publications of the Astronomical*

Hendricus Frederikus van der Kalle was a Dutch writer of detective stories. This quote is from *Lijk Halfstok* (1948) [54]: 'Rommelig', besloot Charles C. M. Carlier. 'En slordig!' Het was, als je 't goed naging, eigenlijk allemaal nogal rommelig, slordig eigenlijk, en onafgewerkt; –de schepping en zo, de sterrenhemel en zo, zelfs al was het de zomerse sterrenhemel van zijn geliefd Provence. 'Typisch haastwerk', zei mijnheer Charles C. M. Carlier.

M.C. Escher was a Dutch graphic artist, often inspired by mathematical concepts and regularities, see [55]. As the trema shows the name Boötes if pronounced with three syllables

© The Editor(s) (if applicable) and The Author(s), under exclusive license
to Springer Nature Switzerland AG 2021
P. C. van der Kruit, *Pioneer of Galactic Astronomy: A Biography of Jacobus C. Kapteyn*,
Springer Biographies, https://doi.org/10.1007/978-3-030-55423-1_6

Laboratory at Groningen, in which the results of the astronomical work in Groningen were published. Apparently there was a backlog to make up for, because in the first year no less than eight parts appeared. After he was appointed member of the Royal Netherlands Academy of Arts and Sciences (KNAW) in 1888, he also published regularly in the minutes of its meetings. Kapteyn was a very active and loyal KNAW member, who almost always attended the meetings. These took place on Saturdays (in Amsterdam) and were followed by an informal dinner, after which he still had to make the return trip to Groningen. In the meantime, after using different temporary buildings, he had finally gotten his own laboratory.

6.1   The Astronomical Laboratory

When Kapteyn started measuring the photographic plates for the *CPD*, he was allowed the use of two rooms in the Physiological Laboratory of his friend Dirk Huizinga to set up his measuring equipment. This was located diagonally behind the Academy Building. A large part of the measurements were done here and he also put up there his second parallactic instrument, which the *Carte du Ciel* had asked him to build as a prototype. That was not an ideal solution at all. Despite the limited space, Kapteyn already referred to his work-space as the 'Astronomical Laboratory'.

In 1881, he had submitted a detailed plan for his own observatory, but that came to naught. He had even identified a location to the south of the city of Groningen (on the site of the current Tax Administration offices at the Kemkersberg), from where there was an unobstructed view of the southern sky; one looked in that direction towards the training grounds belonging to the barracks located there (the current Sterrenbos and Dr. S. van Mesdag Forensic Psychiatric Center). A design had already been made for the building. In 1890 he adapted the plan and presented a new design, somewhat more modest in size, to house a photographic telescope. Not long after that, during academic year 1890–1891, it was his turn to become Rector Magnificus. In his scientific lecture on the occasion of the transfer of the rectorate at the end of his tenure, entitled *The importance of photography for the study of the higher parts of the heavens*[1], he elaborated extensively on the importance of photographic studies in astronomy and the fact that there was no photographic telescope in the Netherlands. By the 'higher parts of the heavens' he meant that part which was not around the ecliptic; after all, most objects in the Solar System (certainly the planets and most of the asteroids) were limited in their positions in the

---

[1] De beteekenis der photographie voor de studie van de hoogere delen des hemels.

sky to a band not too far from the ecliptic. The 'higher parts' then refers the terrain of the fixed stars.

Kapteyn stated that there would be a good division of tasks in astronomy in the Netherlands, if Leiden limited itself to astrometry, and left astrophotography to Groningen. Utrecht then would do its own specialties. Van de Sande Bakhuyzen, however, felt otherwise. At the start of the *Carte du Ciel* project, he had initially announced that Leiden could not afford the purchase of a telescope to participate fully in the project. The costs were estimated at 50,000 guilders (about 650,000 € in current purchasing power). But around 1890 he finally realized the great importance of photography for astronomy. Kapteyn's plans probably played a role in this and he wanted to prevent such an important instrument from going to Groningen and demoting his observatory to second place in the country. Of course he had the advantage of existing infrastructure; the designs for the buildings for a photographic telescope for Leiden and Groningen, both made by the Government's 'Chief Architect' Jacobus van Lokhorst (1844–1906), show a much simpler and cheaper building for Leiden. As a result, the choice for the Minister was not too difficult; probably for a large part on that basis Leiden got the photographic telescope. The instrument was in the Ministry's budget for 1895.

In his speech as retiring rector in 1891 Kapteyn referred to the situation that there was an observatory in the Cape of Good Hope, where there was too much work to do at the telescopes to leave time and manpower for the processing of the photographic plates, while in Groningen it was the other way around. After the selection of Leiden for the photographic telescope, he must have given up his plans for his own observatory. But, he changed his plans aiming for a well-equipped laboratory to perform measurements on plates that were taken elsewhere, where the manpower and instruments were lacking to do the job. In 1892 he submitted a request to purchase a third instrument in addition to the measuring devices for the *CPD* and the *Carte du Ciel*: 'When another measuring device from Repsold for about 2400 f. is obtained, we will have here an astronomical facility here with possession of a unique collection of photographic measuring instruments [...] and we will therefore have the opportunity here in Groningen, better than anywhere else, even abroad, to study the various methods and measurements of photographic images of star fields'. The amount of 2400 guilders corresponds to about 30,000 € purchasing power nowadays.

In 1896 Kapteyn was finally able to inaugurate his Astronomical Laboratory. He was given the use of the official residence of the Queen's Commissioner[2].

---

[2]This was Queen Emma, Adelheid Emma Wilhelmina Theresia, Prinsess von Waldeck und Pyrmont (1858–1934), regent for her daughter the later Queen Wilhelmina.

The commissioner at that time, Carel Coenraad Geertsema (1843–1928), already lived in Groningen, so he did not use the official residence. The building, at the address Oude Boteringestraat 44—the current offices of the University's Governing Board; see Fig. 4.7 C—was only a temporary home for the laboratory, and no reconstruction work was allowed to facilitate the measuring instruments and make them stable and free of vibrations, but still, it was Kapteyn's own laboratory. At the official convocation on January 16, 1896, Kapteyn gave a public lecture in which he described the role of his 'observatory without telescopes' as complementary to that of the other observatories: measuring and interpreting photographic data from elsewhere. Indeed, the fact that Kapteyn never got his own observatory can be seen as a blessing in disguise. Had Kapteyn been responsible for the operation of the Dutch photographic telescope (which did not produce all that many prominent results in Leiden), the discoveries that he was still to make and the very prominent international role Kapteyn was to play would probably never have come about, or at best to a much lesser extent.

When in 1904 the building had to be used again for its original purpose, Kapteyn was housed in a building on the corner of the Oude Boteringestraat and the Spilsluizen (see Fig. 4.7, D). Originally it was the home of the military guard that defended the town; today it is home to the 'Hotel Corps de Garde'. This solution, too, was temporary. But in 1911 the Astronomical Laboratory finally got its own permanent housing. Ironically, it was the same building in which Dirk Huizinga had initially accommodated him. Huizinga had died in 1903 and was succeeded by Hartog Jacob Hamburger (1859–1924), whom we will meet again later. Hamburger moved to a larger building and Kapteyn was given the use of the vacated building (see Fig. 6.1) behind the Academy Building.

The layout of the building is shown in Fig. 6.2. This map is taken from a contribution by Kapteyn to a commemorative volume published on the occasion of the University's three-hundredth anniversary in 1914. Kapteyn's room was number 10 on the first floor, his assistant was in room 9 and the library was in room 15. The other rooms upstairs were for the (human) calculators. The ground floor was for measuring equipment and the rooms there had a separate foundation for stability. The rooms with the numbers 2 and 6 had been for Kapteyn's use at the time Huizinga had made them available for the *CPD*. One of these rooms can be seen in the picture in Fig. 6.3, which comes from the same publication from 1914. Later, in the 1930s, Kapteyn's successor Pieter Johannes van Rhijn obtained a telescope on top of this building, but it was removed at the end of the 1950s by his successor Adriaan Blaauw (1914–2010). In the 1980s a part of the upper floor was damaged in a fire; the university has (unfortunately) seized the opportunity to have the building demolished.

Fig. 6.1 The Physiological Laboratory after Kapteyn had moved in and it had become the Astronomical Laboratory. Kapteyn Astronomical Institute, University of Groningen

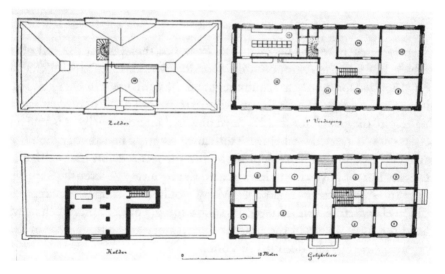

Fig. 6.2 Floor plan of the Astronomical Laboratory (formerly the Physiological Laboratory). University of Groningen

Fig. 6.3 Two rooms with Kapteyn's measuring instruments in the Physiological Laboratory. The rear one was used for the *CPD*. University of Groningen

Kapteyn may have had his own laboratory, but this was not the end of his troubles. In 1907 the Astronomical Laboratory was threatened with closure. He had applied for money for equipment and the parliament in Den Haag saw this as an opportunity to raise the question of the raison d'être of the laboratory. This despite the great fame that Kapteyn now had internationally. It is true that this danger was averted—albeit with difficulty; Kapteyn had already mobilized David Gill to submit a flaming protest—but the fact remains that neither his university nor the government have treated Kapteyn very generously. Kapteyn's loyalty to Groningen is remarkable. Many would have sought or accepted a position elsewhere. Kapteyn never did such a thing. For example, when he was offered the directorship of the Utrecht Observatory at the retirement of Jean Oudemans in 1897, he declined this offer.

6.2   Stars and Galaxies

In Chap. 3 I have summarized the state of astronomy around 1875. For a good understanding of Kapteyn's further work, it is also necessary to place this in the context of what we now know about the Universe outside our Solar System.

As far as stars are concerned, one has to remember that these occur in a great variety. First of all, they can have different intrinsic luminosities. I usually use the word 'luminosity' for that, but also sometimes absolute magnitude. This is the apparent magnitude a star would have at a standard distance. We saw above that the parallax is measured using the projection in the sky of the Earth's orbit around the Sun. This generally gives an ellipse, of which the semi-major axis is called the parallax. If this is 1 second of arc, the distance is defined as 1 parsec (pc). The absolute magnitude is then the apparent magnitude a star would have if the distance were 10 pc (and the parallax 0″.1).

Stars can also have very different masses, ranging from about a tenth to several tens of that of the Sun. A star spends most of its life as a so-called Main Sequence star. This sequence was first noticed in the Hertzsprung–Russell diagram (see Fig. 6.4), which Ejnar Hertzsprung, the later son-in-law of Kapteyn, and the American Henri Norris Russell (1877–1957) independently devised around 1910 (for more background, see Appendix A.5). The diagram shows that for the vast majority of stars there is a fundamental relationship between their luminosity (the vertical axis of the diagram) and the temperature at the surface (the horizontal axis). This feature has been called the Main Sequence. The temperature at the surface of a star includes relatively much blue light when the star is hot and red light when relatively cool. For the horizontal axis of the diagram the spectral type can also be used. In cool stars we see different dark spectral lines of elements than we do in hotter stars. These lines are created by the absorption of light from the star by atoms in the outer layers of that star. Particularly the temperature, but also other physical conditions in these external parts determine which chemical elements and associated spectral lines will be visible. It turns out that along the Main Sequence the mass of the stars also changes systematically: massive stars are blue and hot and emit much light; light stars are red, cool and faint. Our Sun is in the middle of the Main Sequence (Fig. 6.4).

During this Main Sequence phase of its life, in which our Sun is now, the temperature in the central parts of a star is so high that nuclear reactions can occur, because the energy in the motions of the atoms is so great that they can overcome the repulsive force of electrically positively charged nuclei. During these reactions hydrogen is converted into helium and energy is released. This longest phase in the life of a star lasts for different times according to the mass of a star. For a star like the Sun, it lasts about ten billion years and the temperature at the surface (as deep as we can see from outside into the Sun) is about 5800 Kelvin. A star that is much heavier than the Sun is much more wasteful with its hydrogen; the process is then much faster and the temperature at the surface much higher. A star with a mass of one tenth of the Sun stays on the Main

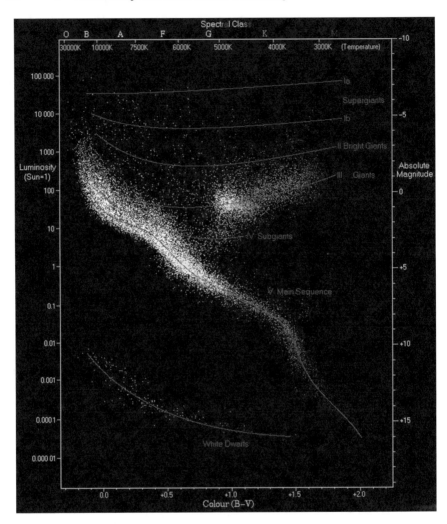

Fig. 6.4 Hertzsprung–Russell diagram of stars. The horizontal axis represents the color or temperature of the surface of the star, the vertical axis the total luminosity or amount of energy emitted in the form of light. For more details see Appendix A.5. Most stars are in the band from top-left to bottom-right, which is called the Main Sequence. From *An Atlas of the Universe* [56]

Sequence longer than the current age of the Universe and is much cooler at the surface. For even lighter stars, the original contraction never makes the inner parts hot enough to trigger nuclear reactions

When the hydrogen in the inner parts has been used up, the star expands and cools down. In this phase the star is a red giant. But the inner parts are contracting and getting hotter. For a star like the Sun, this means that eventually

Fig. 6.5 A panoramic view of the Milky Way. This view of the full $360°$, made up of many separate photographs, has been produced by the European Southern Observatory ESO. The two spots to the right the Milky Way above of the center are the Magellanic Clouds, companions to our Galaxy. The two just to the right of the Milky Way at the bottom are the star cluster the Pleiades (extreme left) (see also Fig. 9.12) and the nearest spiral galaxy, the Andromeda Nebula (see also Fig. 6.7). The blob just to the left of the Milky Way to the extreme top is the constellation Orion with the Orion Nebula (see also Fig. 9.11). Other bright objects are planets that happened to be in the field when the pictures were taken. European Southern Observatory [57]

new nuclear reactions can occur, whereby carbon and oxygen are produced from the helium. That phase does not take very long and eventually the star cools down; what remains is a very compact so-called white dwarf. However, if the star is significantly more massive than the Sun, further contraction will occur once most of the helium has been converted into carbon and oxygen, which will result in even higher temperatures and further nuclear reactions up to chemical elements like iron. Eventually the inner part will collapse and cause a huge shock wave in the star, which then explodes as a supernova. In a short period of time, it will be billions of times brighter than the Sun. More nuclear reactions occur, in which all possible chemical elements up to the most massive like uranium are formed and thrown into space. What remains is a very compact neutron star or black hole.

A galaxy like ours (the Milky Way Galaxy; see Fig. 6.5) consists roughly of two parts when it comes to the stars. One part is called disk and the stars in it are referred to as disk population. They form a flattened structure, because it rotates around the center. The Sun lies about 8000 parsec (25,000 light years) from the center of our Galaxy (about halfway its radius); the rotation speed here is about 220 km/s. The random relative velocities of the stars are much smaller, of the order of a few tens of km/s. The second component of the system is called halo and the stars in it form the halo population. The most conspicuous structures in it are the globular clusters, on average consisting of one hundred thousand stars. This division was not yet known in Kapteyn's

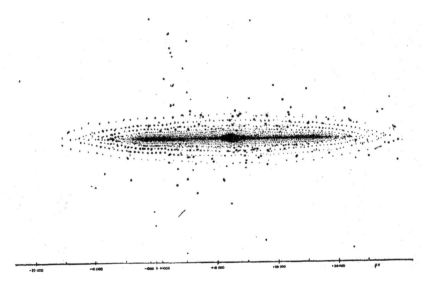

Fig. 6.6 Our Galaxy according to Jan Hendrik Oort. This figure comes from a book by Willem de Sitter, *Kosmos*, published in 1934

time; Fig. 6.6 shows a sketch by the famous Dutch astronomer Jan Hendrik Oort (1900–1992), made in the twenties or thirties and published in a book by Willem de Sitter (1872–1934). Jan Oort and Willem de Sitter were both students of Kapteyn. The halo as a whole rotates slowly, but stars have high velocities among themselves.

The halo stars are all as old as the Galaxy itself, about ten billion years. But in the disk there is also gas and dust between the stars, from which new stars can form. So in the disk there are very young but also very old stars. There is another fundamental difference between the halo and the disk, which only became apparent in the middle of the twentieth century. When the Universe was three minutes old, it was almost exclusively composed of hydrogen (for three quarters) and helium (for one quarter). The same goes for the first stars. But the heavy stars among them evolved rapidly and became supernovae, adding other chemical elements to the gas between the stars (especially carbon, nitrogen and oxygen, but also all the others, in sometimes very small amounts). Later generations of stars therefore contain more chemical elements that are heavier than helium (astronomers refer to this with the term abundance); for the Sun this is only 2% of the mass in total. This is typical for the disk population. The stars in the halo have abundances that are usually between ten and a

Fig. 6.7 Two galaxies similar to ours. On the left is the Andromeda Nebula or M31, which is the nearest large galaxy. On the right a system where the disk is seen edge-on, known as NGC 891 [58, 59]

hundred times smaller than that of the Sun. The formation of the chemical elements in stars from hydrogen and helium is called stellar nucleosynthesis. The relative amount of the various chemical elements in the Universe can be understood in detail by this process, and I consider this to be one of the greatest achievements of science.

In addition to our Galaxy, there are many other galaxies. Some look a bit like ours, like the ones in Fig. 6.7, but they can also look very different. This is because the relative contribution of the halo and the disk to the system may be very different. For a good understanding of Kapteyn and his work, there is no need to go into this in more detail here.[3]

6.3   Parallaxes and Proper Motions

We saw above how Kapteyn showed that parallaxes of stars could be measured from the timing of their passing the meridian. This success also had a downside, because it was now clear that measuring parallaxes on a large scale was at least as difficult as feared, and that the determination of distances of large quantities of stars this way was impossible. Of course these measurements can also be done photographically (with three exposures half a year apart) and Kapteyn even suggested to do this as a matter of routine in the *Carte du Ciel*. But that, as we have seen, was not accepted. The method is basically simple. Take a picture of the sky when the shift of a star through the annual parallax is maximal. Leave the plate for half a year until the shift is maximal in the opposite direction, and make another exposure of the same star field on the same plate, but now with the center of the plate shifted a little bit. Repeat that after another six months. Assume that the faint stars on the plate are predominantly far away and have little parallactic motion of their own and measure the shifts relative to those background stars. The difference between the first and third recording then show the annual proper motion. The second recording shows, compared to the mean of the first and the third, the annual shift due to the parallax.

Kapteyn's experience with the *Carte du Ciel* had, despite the decision not to use his parallactic measuring method, also a positive outcome. At the great congress in Paris in 1887 he had met Anders Severin Donner (1854–1938) of the Helsingfors Observatory (Helsingfors is Swedish for Helsinki). They were among the younger participants in the congress (see Figs. 5.10 and 6.8) and between them a friendship developed that would last their whole lives. Anders Donner was in favor of Kapteyn's idea to measure parallaxes and proper motions and with his telescope at Helsingfors—which he acquired for his participation

---

[3] For more background see for example my course *Structure and dynamics of galaxies* [60].

Fig. 6.8 Anders Severin Donner (1854–1938) of Helsingfors Observatory. This picture comes from the *Album Amicorum*, presented to H.G. van de Sande Bakhuyzen on the occasion of his retirement as professor of astronomy and director of the Sterrewacht Leiden in 1908 [47]

in the *Carte du Ciel*—he collected an enormous amount of photographic material for Kapteyn over time. In doing so, he followed the method Kapteyn had suggested of three exposures on each plate with half-year intervals. The first publication in the series mentioned above, *Publications of the Astronomical Laboratory at Groningen*, concerned such a study, and it must have been finished some time before it was published in 1900. In that study an attempt was made to measure the parallax of 248 stars on plates, which covered a small part of the sky. The analysis showed that an accuracy for parallax measurements of $0''.02$ could be achieved for a combination of three such plates of the same area. Not more than ten stars were found to have a measurable parallax. Until 1925, 39 volumes appeared in the *Publications of the Astronomical Laboratory at Groningen*, of which no fewer than nine were based on plates taken by Anders Donner.

Kapteyn's main program was the study of the distribution of the stars in space, but sometimes other things came his way. There was the discovery of Kapteyn's star. This was a by-product of the *CPD* measurements. The star had

Fig. 6.9 The proper motion of Kapteyn' star in the sky between 1975 (top), 1990 (middle) and 1996 (below). This figure is produced by the author from public, digitized *Sky Surveys* [61]

been photographed in 1890 and 1893 for the *CPD*, and the relevant part of the sky had also been mapped out as part of what was to become the *Córdoba Durchmusterung*. But Kapteyn could not find the star in these data. At the Cape, David Gill's assistant Robert Thorburn Ayton Innes (1861–1933) had noticed that in a nearby position a bright star had been seen at Córdoba in 1873; it therefore seemed to be a case of very large proper motion. Observations by Innes at the Cape Observatory in 1897 confirmed this; the star moved in the sky at an unprecedented speed of almost 9 arcseconds per year. Fig. 6.9 shows photographs from more recent years, which clearly illustrate the displacement.

Kapteyn published this discovery also on behalf of the astronomers at the Cape, mentioning Innes in particular, in the *Astronomische Nachrichten*. But not long after that, the British magazine *the Observatory* accused Kapteyn of flaunting the feathers of the 'real' discoverer of the star, Innes. This brought out a less pleasant aspect of Kapteyn's character; he was much offended by this and wrote a strong letter to David Gill. Eventually Gill published a detailed explanation of how this had developed, and pointed out that Kapteyn had indeed carefully mentioned Innes. However, the editors of the magazine stuck to their opinion. Nevertheless, the star was named after Kapteyn, because he published its discovery (albeit on behalf of others).

David Gill raised another matter in his correspondence with Kapteyn about this case. In a discussion of the *CPD* Karl Hermann Gustav Müller (1851–1925), the director of the Astrophysikalisches Observatorium in Potsdam, wrongly claimed that the *CPD*—and even the idea for it—had sprung from Kapteyn's brain. Gill had not made a point of this at the time, nor had he protested to Müller. But he now suggested that Kapteyn could have objected publicly or by letter to Müller and set the record straight. Kapteyn admitted this and wrote extensively about Gill's participation in the project (but without mentioning Gill by name) in the Preface to the last part of the *CPD*. The correspondence concerning this episode shows that Gill was the most tactful and the wisest of the two.

However, they remained excellent friends. Gill visited Kapteyn several times in Groningen, for the first time just before the great *Carte du Ciel* congress in Paris in 1887. In her biography [1] his daughter wrote about this first visit:

Mother looked at them with joy, as she was standing in front of the house when they left arm in arm for the Laboratory (the two small rooms had this grandiose name in the Kapteyn family). Gill talking loudly and gesticulating, stared at by the Groningen citizens, Kapteyn, short and modest, quietly happy next to him. This first visit was a great success, and also the children were delighted with this big playmate, who had won their hearts notwithstanding his incomprehensible language. From then on he was 'Oom Gill' [Uncle Gill] and remained that forever.

Kapteyn visited Gill only once in South Africa, but did visit him and his wife more often when they had moved to London after his early retirement. The friendship with Anders Donner was just as close; Donner hated writing and travelling, so contacts were limited to an absolute minimum. Anders Donner has visited Groningen a couple of times, but as far as I know Kapteyn has never been to Helsinki.

About Kapteyn's star the following should be noted. It later turned out to be one of the closest stars. It is at 12.8 light years from Earth; only 24 other stars (or star systems) are closer than that. It is a halo star with a very low chemical abundance (about ten times less than the Sun) and is at least 10 billion years old. It describes an orbit in space in which it crosses the disk of our Galaxy in a direction opposite to the rotation, so that the velocity relative to the Sun is almost 300 km/s. Only the later discovered Barnard's star, discovered in 1916 by American astronomer Edward Emerson Barnard (1857–1923), has a larger proper motion, but it is only 6 light years away.

It has recently been suggested that the Kapteyn's star is part of a group of stars that may have escaped from the large globular cluster ω Centauri. This is probably not a 'real' globular cluster, in the sense of representative of the halo-population, but a remnant of a small stellar system that has been 'swallowed' by our Milky Way System, during which the outer stars were detached from it. Very recently, two planets have also been found around Kapteyn's star; this was not expected because the star, and hence the gas from which it was formed, contained almost no chemical elements except hydrogen and helium.

Another 'excursion' of Kapteyn concerned Nova Persei 1901. This star suddenly appeared in February 1901 and became one of the brightest in the firmament. We now know that it is an example of a late stage in the evolution of a star; it is a binary star, one component of which is a white dwarf. In such so-called cataclysmic binaries, gas flows at great velocity to the white dwarf when the other component in its evolution becomes a red giant and expands; this releases energy. Of course this was not known at the time; novae had been seen before, but never so bright. So Nova Persei was probably relatively close. It slowly became fainter until in 1904 it could only be seen with the largest telescopes. After a year, a nebula was discovered around the star and a little later it was discovered that this nebula was expanding (Fig. 6.10).

The expansion was so rapid that it had to indicate a small distance. For the following we have to realize that today we all grew up with Albert Einsteins (1879–1955) notion that nothing can move faster than light; however, the special theory of relativity only dates back to 1905. In the case of Nova Persei, it was clear that its expansion rate had to be greater than the speed of light unless the star was very close. But why this occurred was not clear at all. Kapteyn suggested in an article that what we see is a long, thin string of matter and dust, more or less pointing away from the star, along which we see the expanding light-front resulting from the sudden increase and peak in brightness due to the short burst. And that motion then had to be with the speed of light as the light-front traveled along the filament. That was a clever idea and gave a distance of the nova of 91 parsec (Kapteyn did not use that term and spoke of a parallax of 0″.011).

We now know that the distance to Nova Persei is no less than 460 pc, so the apparent expansion must be faster than the speed of light after all. How is this possible? Well, according to Einstein, material things cannot move faster than light. However, this does not apply to a flare of light. Imagine a lighthouse

Fig. 6.10 At the top two exposures of Nova Persei of November 1901 (left) and February 1902 (right). Both are exposures of about 10 hours divided over a few nights. This was done at Lick Observatory of the University of California. Below a detailed view of the same area after some image processing. Tony Misch, director of the Lick Observatory Historical Collections re-scanned the original plates for me

and a long screen at a great distance. The further away that screen is from the lighthouse, the faster the spot of light moves. Nothing fundamental stops us from moving the screen so far out that the spot moves faster than light. The light itself reaches the screen with the speed of light, but the spot can in principle move faster. It now turns out that the basis of Kapteyn's idea, the motion of a light-front through interstellar dust, was correct. He is therefore regularly mentioned as the founder of this general explanation of 'superluminal' speed.

6.4 Willem de Sitter

Due to the lack of observing facilities, Kapteyn had few students. He could provide them after all with very little experience with observing. However, there were students in mathematics and sometimes physics, who did some practical work with him in the form of measuring of plates. In principle they were not reimbursed for this, but it was some good experience. In 1896 Willem de Sitter was one of those students. Jan Guichelaar has published an excellent biography of Willem de Sitter in this series [62]. De Sitter was born in Sneek (Friesland) in a family that mainly consisted of lawyers, but Willem broke with this tradition and had come to Groningen to study mathematics. He attended Kapteyn's lectures and worked in Kapteyn's laboratory when David Gill visited Groningen. Later, in a letter to Gill, de Sitter mentioned that this took place on October 2, 1896; he also mentioned his poor English, of which he was still embarrassed. But with Kapteyn's help—and when he had to lecture that of Mrs Kapteyn—they had a conversation in which Gill became much impressed by Willem de Sitter. So impressed in fact, that the next morning he asked de Sitter to come and see him and offered him to come to Cape Town to do work and prepare an astronomical PhD thesis.

De Sitter (see Fig. 6.11) did not have the intention at all of obtaining a PhD in astronomy; he had come to Groningen to study mathematics. But after

Fig. 6.11 Willem de Sitter. This picture comes from the *Album Amicorum*, presented to H.G. van de Sande Bakhuyzen on the occasion of his retirement as professor of astronomy and director of the Sterrewacht Leiden in 1908 [47]

consulting his parents, he decided to accept the offer. After passing his doctoral exam (masters), de Sitter indeed traveled to Cape Town in August 1897. He was supposed to make photometric observations of stars in the *CPD*. For this he had brought a Zöllner photometer (see Fig. 2.6). However, the telescope he was to work with had not been completed in time. He did make several observations for Kapteyn's program, but spent most of his time observing the four large satellites of Jupiter. This concerned precise measurements of the positions to study the characteristics of their orbits.

The satellites or moons of Jupiter had been discovered by Galileo Galilei when he was the first to look at the sky with a telescope (and report on it in a publication) in 1609 and 1610. Now their orbits were fairly well known, but there were also mutual attractions that ensured that there was systematics in these orbits in the form of resonances, so that they circled around Jupiter in a fixed regularity. The inner three moons (Io, Europa and Ganymede) do so in orbital periods that have almost exactly the ratio 4 : 2 : 1. Think of a situation where the inner two satellites pass each other in their orbits. If Io (who is closest to Jupiter) has gone around twice and has returned to the same position, Europe also has returned there after one orbital period. If the ratio between their periods is not exactly 1 : 2—for example if Io at that time would be slightly ahead of Europa—their mutual attraction will slow down Io a bit and accelerate Europa somewhat, changing that ratio back closer to the exact ratio 1 : 2. In this way they get stuck in a stable pattern of mutual perturbations that returns regularly after one or more orbital periods.[4] This is much more common in the Solar System, also among the planets; for example, Jupiter and Saturn are close to a 5 : 2 resonance and the Earth and Venus in a 3 : 8 one. This makes the Planetary System rather robust. In practice, the situation is more complicated, because Jupiter's satellites orbit only in approximately circular orbits around the planet and also in slightly different orbital planes. In the case of Jupiter there also is a fourth satellite, Callisto, which is in resonance with Ganymede of 3 : 8.

After his return to the Netherlands in December 1899—he had married Eleonora Suermondt (1870–1952) in 1898—de Sitter wrote his dissertation on the subject of the orbits of the Jupiter satellites; he was Kapteyn's first PhD student when he obtained his degree in 1901. This issue of the Jupiter satellites has interested him for the rest his life and he did much work on it until the end. He became much better known among a wider audience because of his work on Einstein's general theory of relativity and their joint publication, which

---

[4]If the satellites were to orbit in pure circles in one plane and if there were only two them, the orbital times would be in exact resonance, but in the actual situation it will be approximate. A regular pattern is already sufficient to create and strengthen the orbital resonances in an approximate manner.

introduced the Einstein–de Sitter model of the Universe. However, de Sitter himself seems to have considered his work on Jupiter's satellite system as more significant.

De Sitter was next appointed as Kapteyn's assistant. He started with what he had actually gone for to the Cape. Kapteyn had noted that on the photographic plates for the *CPD* not only relatively more stars could be seen in the Milky Way than at higher Galactic latitudes, but that this effect was even more pronounced than in the Durchmusterungs made visually and with meridian circles (the southern extensions of the *Bonner Durchmusterung*, including those of the Córdoba Observatorio Astronómico in Argentinië, which had been started in the meantime). Because photographic plates are more sensitive to blue light than the eye, this would mean that there should be relatively more blue stars in the Milky Way. Kapteyn had already presented an extensive study on this subject for the KNAW in 1892. One approach would be to use new photographic emulsions (so-called isochromatic emulsions), which had a color-response more similar to that of the human eye. Gill was already using these to take plates in areas far away from the Milky Way. De Sitter then measured these up. The result was published in 1900 as a second contribution to the *Publications of the Astronomical Laboratory at Groningen*. The conclusion was not very conclusive; it seemed that there were systematic errors in the visual estimated magnitudes. De Sitter then investigated this in detail using more photometric and visual observations done at the Cape. In the meantime Robert Innes had obtained more of such observations, that de Sitter had not been able to do because the weather had been bad during his stay there. In the Volume XII of the *Publications of the Astronomical Laboratory at Groningen* in 1904, he confirmed Kapteyn's findings, but also showed that the matter was actually more complicated because it appeared that the number of blue stars fluctuated at least as much along the Milky Way as it did perpendicular to it.

As Kapteyn's assistant, de Sitter was closely involved in measuring the plates Anders Donner had obtained in Helsingfors to measure parallaxes and proper motions. Between 1902 and 1908, when Willem de Sitter left for Leiden, the *Publications of the Astronomical Laboratory at Groningen* contained four large volumes about such observations in which Willem de Sitter had been involved.

6.5 Secular Parallax

While the production of the *CPD* was still in progress, Kapteyn already took an extensive interest in determining the distribution of the stars in space. We are in the fortunate circumstance that he was asked in 1907 by David Gill to write an overview of his work. Gill had taken early retirement when he was 63 and had moved back to the United Kingdom, where he had settled in London. In 1907 he was chairman of the 'British Association', a society of scholars; he wanted to use his presidential address later that year in Leicester to present a discussion of Kapteyn's work. The manuscript of the document that Kapteyn wrote for Gill is in the Groningen archives [63]. It is a very nice, systematic summary of how Kapteyn had developed his work, and well worth reading. In the rest of this section I follow Kapteyn's own description of his work.

Kapteyn had already realized that the direct measurement of parallaxes for this program was not only time-consuming, but that this way only the distances to the nearest stars could be measured. Therefore he turned to statistical methods to measure stellar distances, using the motion of the Sun relative to the stars in its neighborhood. This gives a pattern in the sky of proper motions pointing away from the point towards which the Sun moves, the Apex (see above), and to the opposite point, the Antapex. If the direction and magnitude of the Sun's motion through space were known and if all stars would stand still, their proper motion would be a direct measure of their distance. Figure 6.12 shows the principle. The Sun has a velocity X and the angles $a \to b$, etc. are the proper motions of the stars that have velocities $A \to B$, etc. as reflections of

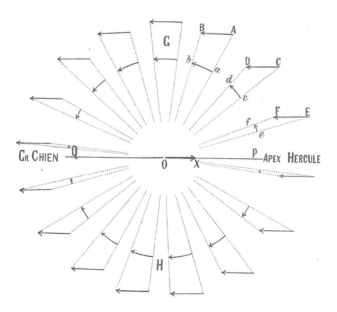

Fig. 6.12 Illustration of the principle of statistical measurement of parallax by Kapteyn in a lecture (in French) in 1906. The Apex is on the right in the constellation Hercules and the Antapex on the left in Canis Major. Kapteyn Astronomical Institute, University of Groningen

that of the Sun, pointing away from the Apex. If the star is farther away, the angle $a \rightarrow b$ will be smaller and the parallax can be determined from this. In practice, stars themselves also move in space, so this can only be done using a statistical approach. For example, one can do this for a collection of stars of which there is reason to assume that their distances are similar. The result then is called the 'secular parallax'.

The method to determine the position of the Apex from measurements of the proper motions of groups of stars had already been worked out in 1843 by the Frenchman Auguste Bravais (1811–1863). Kapteyn started out with a new and more accurate determination of the position of the Apex. As I described above, the problem arose that this determination depended on the precession of the equinox, the motion of the Earth's axis in space. For the determination of proper motions, two measurements at different times are required, and in order to compare positions measured at different times you need to know how the position of the Earth's axis in space has changed in the meantime. Kapteyn used the available data of the proper motions of the 'Bradley's stars'— as Kapteyn referred to them—, the stars from the catalog of James Bradley and Friedrich Wilhelm Bessel (see Sect. 3.2). Using this catalog as a basis, Georg Friedrich Julius Arthur von Auwers (1838–1915) had produced a fundamental catalog[5] in 1881, based on observations in Greenwich and Berlin; it contained the proper motions of 3268 of those 'Auwers-Bradley' stars. Kapteyn took different assumptions for the position of the Apex, for the precession and for the unknown declination errors in the Bradley catalog to study the effect these assumptions had on the derived position of the Apex. He published this enormous amount of arithmetic in 1900 and in 1902 in the *Publications of the Astronomical Laboratory at Groningen*. Fig. 6.13 shows his determination of the Apex and the current determination; Kapteyn's error was less than 10°!

When the Apex was known with sufficient accuracy, one still had to determine the space velocity of the Sun to sufficient accuracy to apply the method of the secular parallax. Kapteyn used measurements of radial velocities (along the line of sight) of 51 stars, obtained in Potsdam by Paul Friedrich Ferdinand Kempf (1856–1920). If you know roughly where the Apex is, you also know how much of the velocity of the Sun is along the line of sight; by assuming that the *average* velocity of the stars themselves is zero, you can make an estimate of the velocity of the Sun relative to those stars. Kapteyn, partly with the help of independent determinations by others, eventually arrived at a value of 19 km/s. That means that the Sun travels in one year a distance in space equal to four times that between the Earth and the Sun; so with the secular parallax method

---

[5]'Fundamental' here means that the positions and proper motion were measured absolutely and not relative to other stars.

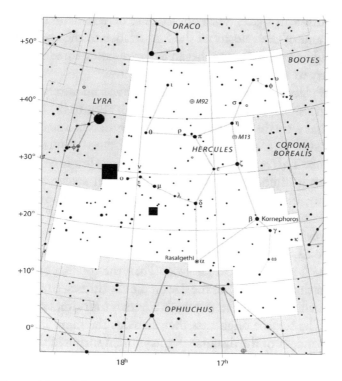

Fig. 6.13 The position of the Apex of the Sun on the map of the constellation Hercules (with Lyra next to it), from the website of the International Astronomical Union [64]. The large, black square indicates Kapteyn's determination of the position of the Apex; the small square the current one. The bright star Vega is the black round dot in Lyra

you have a basis for distance measurements that after one year is already four times as large as that for the direct parallax, which after all is the radius of the Earth's orbit. This argument by Kapteyn is a bit misleading; after all one measures to full major axis of the parallax on the sky, which corresponds to the *diameter* of the orbit of the Earth. If you now know the velocity of the Sun, you can correct the radial velocities of stars for that, and so Kapteyn found that the average velocity of stars through space is almost twice (1.86 to be exact) that of the Sun.

This was the beginning of what would later be called *statistical astronomy*, of which Kapteyn then was one of the founders.

6.6 Spatial Distribution of Stars

Statistical astronomy, however, involved more than this. It is first and foremost the determination of properties of stars by statistical methods, in particular their distribution in space. With an accurate determination of the position of the Apex and of the velocity of the Sun relative to the nearby stars, a first attempt could be made. Kapteyn began by formulating three assumptions that were necessary to obtain an unambiguous result.

In the *first* place Kapteyn assumed that there was no dust between the stars, at least not to a significant extent, that scatters or absorbs the light of the stars. After all, if that were the case, the apparent brightness of a star would not only be determined by its intrinsic luminosity and distance. Kapteyn figured out a way to test this. This requires the concept of surface brightness. In astronomy nowadays this is often expressed in magnitudes per square arcsecond. For a dark night sky, in visual light (visible to the eye), that is something like 22 or 23; that means that for an area of one square second of arc (one by one arcsecond), the amount of light that comes from it is the same as that of a star of magnitude 22. Now look at a nebula. It also has a certain surface brightness. Now imagine it to be twice as far away. Then the total amount of light you receive is four times smaller. But the size on the sky (in arcseconds, for example) is twice as small; so the area in the sky it covers is four times smaller in square arcseconds. The surface brightness then remains the same and this property thus is independent of the distance. Kapteyn then surmised that if there were absorption of light in interstellar space, the surface brightness of nebulae at greater distances would have to be reduced on average. But he could not find any unequivocal signs of this.

Secondly, Kapteyn assumed that the motions of the stars in space were random, i.e. that there was no preferred direction and that the average velocity was the same everywhere and in all directions. With these assumptions he was able to estimate the average distance of stars as a function of apparent magnitude from the available observations of proper motions. The result was a table in which, if you look up a particular apparent magnitude and a particular proper motion, you find the average distance.

For individual stars with a measured parallax, you can of course compare this to the result you get from this method. And if you would have enough of such stars, you can also determine how in general the stars are distributed around those averages. This is what Kapteyn called the 'frequency law'. To see how that works, I cite how Kapteyn explained that to David Gill in his resumé from 1907 [63]. If you are not interested in such details, skip to the next paragraph.

Take stars of magnitude 6, so $5^m.5$ to $6^m.5$. There are about 4800 such stars all over the sky. According to Auwers-Bradley [i.e., the catalog], $9\frac{1}{2}$ percent of such stars, i.e., about 460, have proper motions between $0''.04$ and $0''.05$ per year. According to the formula, the average parallax of such stars is almost exactly

0″01. And according to the Frequency Law, 29% of those stars have parallaxes between the average value and double that value; and in the same way 6% have parallaxes between two and three times the average. So of our 460 stars, 133 will have parallaxes between 0″01 and 0″02, 28 between 0″02 and 0″03, and so on. Find distances for other stars of magnitude 6 in this way and then treat stars of the 1st, 2nd, ..., 9th magnitude in the same way and you will eventually have located all these stars in space.

In principle you can determine the distribution of the stars over intrinsic luminosity (i.e., how many there are relative to each other as a function of luminosity), expressed as absolute magnitude. In other words: you can determine how in a certain part of space the stars are distributed in absolute magnitude or luminosity. Kapteyn called this the luminosity curve, nowadays astronomers say the luminosity function. And the *third* assumption that he made was that this function would be the same everywhere in space. What then remains is the total number of stars as a function of the distance from the Sun, the 'density curve'.

With these three assumptions it is in principle possible to derive from star counts and observations of proper motions what this density curve is. In practice this was not so easy, because it required the use of complicated mathematical methods. That turned out to be so difficult, that Kapteyn called upon his brother Willem in Utrecht. Together they succeeded in developing the necessary mathematical tools. This produced the first of two major articles, which became number five in the series *Publications of the Astronomical Laboratory at Groningen*. The second article, the application of these mathematical tools, was supposed to appear as part six, but it never did. There were indications that something in the assumptions was wrong, which Kapteyn first attributed to the poor declinations of the Auwers-Bradley stars. Slowly, however, he came to the realization that he had to drop the assumption of a lack of preferred directions in the spatial motions of the stars. That realization was a set-back but at the same time would be a discovery that would only enhance his already growing fame.

6.7   Natural Sciences Society

An important activity of Kapteyn in Groningen concerns his efforts for the Natural Sciences Society. This association still exists (I have been chairman of the Board between 1998 and 2020) and since 1976 it is 'Royal'. It was founded in 1801, with the aim of promoting a broad knowledge and understanding of natural science [65] . Not only physics, astronomy, chemistry, biology and so

Fig. 6.14 Physiologist Dirk Huizinga. Huizinga was a good friend of Kapteyn and a member of the board of the Natural Sciences Society. He is the father of the famous historian Johan Huizinga. University of Groningen [68]

on were understood as covered by natural sciences, but also medical science was included. A short history of the KNG in Dutch can be found in the bicentennial publication of 2001, *Een spiegel der wetenschap* [66], but see also [67].

Shortly after he arrived in Groningen in 1878, Kapteyn was already giving lectures for this Society. In one of his letters to his fiancée, he wrote that he had been very nervous before giving his first lecture, but that things had gone well. That lecture took place on November 28, 1878 and covered a whole range of subjects. The title was *About the relationship between different stellar systems and more specifically about the clusters of stars, the binary and multiple stars, the Galaxy and the nebulae.* So he treated pretty much everything outside the Solar System. More lectures followed on comets, astronomical accuracy and falling stars; Kapteyn soon became an honorary member because he was considered one of the regular speakers.

In 1886 Kapteyn was invited to become a member of the board. This must have had something to do with the fact that his good friend Dirk Huizinga (see Fig. 6.14), who had allowed him to use rooms in his laboratory to measure the *CPD* plates, was a member of the board. Kapteyn kept lecturing regularly. In 1897, the Society reorganized itself and established, among other things, a 'Central Bureau' that would collect and discuss knowledge about Groningen and its environs. It was chaired by Pieter Roelf (Roelof) Bos (1847–1904), geography teacher at the HBS in Groningen. He is best known from his 'School Atlas of the Whole World', popularly known as the Bosatlas. Kapteyn also became a member of the board of this department. In addition, a 'Scientific Department' was set up on the initiative of Johan Frederik Eijkman (1851–1915), a professor of chemistry. However, when the statutes of this division had to be formally adopted in 1900, Eijkman withdrew, because in his opinion the division did not get enough autonomy from the board. The new chairman of the department was Arnold Frederik Holleman (1859–1953), chemist and director of the National Agricultural Laboratory; Willem de Sitter acted as secretary. Kapteyn also became a member of the board of this department.

In 1904 Holleman went to Amsterdam, where he was appointed professor; Kapteyn took his place as chairman of the Scientific Department. He remained at this post until his retirement in 1921! Because he was also a member of the board of the Central Bureau, he did resign as a member of the overarching board of the Society. Together with secretary Willem de Sitter he organized a program of scientific lectures. After Willem de Sitter had left for Leiden in 1908, his place was taken by Joost Hudig (1880–1967), chemical engineer, mainly interested in biology and connected to the National Agricultural Testing Station. Joost Hudig and Kapteyn were neighbors in the Oosterhaven for some time, between 1908 and 1910 (see Fig. 4.7, 5). In 1929 he moved to the Agricultural University in Wageningen. He became a widower in 1924. Kapteyn's daughter Henriette, author of the Kapteyn biography from which often is quoted in this book, and since 1923 living separated from her husband, Ejnar Hertzsprung, married Joost Hudig in 1937, a few months after she formally divorced from the Danish astronomer.

The Natural Sciences Society, and especially the Scientific Department, flourished under Kapteyn's leadership. One of the founders of the Society, Theodorus van Swinderen (1784–1851), had donated the building Poelestraat 30, the 'Concerthuis' (see Fig. 6.15), to the Society, on the condition that he and his wife were allowed to live there until their deaths. The meetings took place there. In 1928 one of the rooms was renamed 'Kapteyn Hall'. The speakers are

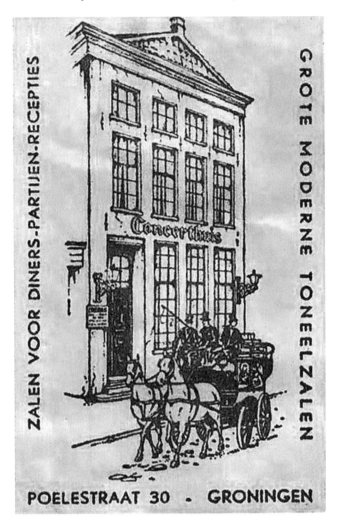

Fig. 6.15 The Concerthuis at 30 Poelestraat, the home of the (now Royal) Natural Sciences Society. This poster dates from the time that Kapteyn was active within the Society. Archives Royal Natural Sciences Society

received there by board members and guests and afterwards there was further discussion between these while enjoying a glass of wine. The room also contains a list of all board members since its founding (Fig. 6.16). Until recently there was a cinema in the building, but in 2012 the rear part, where the cinemas were located, was sold to the municipality of Groningen, which demolished this part and constructed the 'Groninger Forum', a 'general meeting place for residents and visitors of the city of Groningen'. After its completion (2019),

Fig. 6.16 List with members of the Board of the (Royal) Natural Sciences Society at Groningen. This list decorates one of the walls of the Kapteyn Hall in the Concerthuis. At the top the header and in the lower panel the bottom part with Kapteyn on the left on the third line from the bottom (number 47). On the right some of the current board members; in the mean time a second list has been started

the Royal Natural Sciences Society is having its lectures in this Forum and the Kapteyn Hall is used again to receive and entertain speakers.

To further honor Kapteyn's great achievements for the Society, a major public lecture by a scientist with high international standard, the Jacobus C. Kapteyn Lecture, has been held annually since 2012, usually in the auditorium of the Academy Building.

Fig. 7.0 Kapteyn at age 45. This picture is from Henriette Hertzsprung–Kapteyn's biography [1]

7

Star Streams

Everything flows.
Heraclitus (c. 540–c. 480BC)

The wonder is not that the field of the stars is so vast,
but that man has measured it.
Anatole France (1844–1924)

7.1 Gill, Donner, Newcomb

The children were growing up, so by the end of the 1880s the Kapteyns had moved to an upstairs apartment in the Heerestraat (first number S 6a, later 113a; see Fig. 4.7, 4). When exactly they moved there is not known; in 1891 they lived there according to the records of the Groninger Archives, while in 1885 they had moved to the Oosterstraat 42 (see Fig. 4.7, 3). Their third child, Gerrit Jacobus, had been born in 1883 and the need for more space must have been the reason behind the moves. In 1899 they bought a country house in Vries, about 15 km south of Groningen. The house still exists; it now has the address Tynaarlosestraat 58 (in the time of the Kapteyns it was district A 22a) and it bears the name 'De Burcht' (the castle). Whether this name already existed in the Kapteyn period is not clear according to the current owners; in the notarial deeds the name appears for the first time in 1947. It is a detached

---

'Panta rhei', sometimes also translated as 'everything moves'.
Anatole France, born François-Anatole Thibault, was a French poet and novelist. Translation by Alfred Allinson.

© The Editor(s) (if applicable) and The Author(s), under exclusive license
to Springer Nature Switzerland AG 2021
P. C. van der Kruit, *Pioneer of Galactic Astronomy: A Biography of Jacobus C. Kapteyn,*
Springer Biographies, https://doi.org/10.1007/978-3-030-55423-1_7

Fig. 7.1 The second home in Vries where the Kapteyns lived during summers between 1899 and 1909. Photograph by the author with permission from the current owners

house with a spacious garden around it (see Fig. 7.1), just outside the village center of Vries to the east. Further to the east (at about 4 km) lies Tynaarlo, which at that time had a railway station.

David Gill has, as I said, visited Kapteyn a number of times. Henriette Hertzsprung–Kapteyn, in her biography [1], wrote the following about his third visit:

> In 1900 Gill visited Groningen for the third time. He was greeted there as an old tried and tested friend. The Boer War, which brought outrage to the Dutch people, was in its final phase, and the Englishmen were not very popular at the time. Their greedy, unrighteous policies made them hated by the Dutch, who enthusiastically took sides with their fellow tribesmen in the faraway land. The Kapteyn family decided not to bring up this thorny subject, which could become unpleasant for the Brit, and it was never mentioned. Gill's visit was as pleasant and invigorating as ever. His cordial interest and spontaneous, lively spirit were of great influence on Kapteyn's work. Piquant and effective was everything he said and advised, his wisdom a help and comfort in difficult times, his warm friendship

beneficial and his humor irresistible. He allowed himself to be photographed with the two girls, whom he called his 'lassies'[1], teased and spoiled them and was the nicest uncle one could imagine. At one time Gill and Kapteyn visited the physics laboratory in Groningen, where a wise quote was painted above every door. One said: 'Wisdom is better than rubies' (wijsheid is beter dan robijnen). 'I know what that means', Gill exclaimed. 'Whiskey is better than red wine!' He had as much fun himself about his joke as all the others. Typical of Gill was his joking translation of the Latin proverb: Experientia docet. 'You know: Experience does it.'

This passage shows Gill's bond with Kapteyn's daughters. Many later letters ended with a request to send greetings to 'Mrs. Kapteyn' and to the 'lassies'. So what about the son, who was sixteen in 1900 (his birthday was in December) and whom David Gill must have met also during this and earlier visits? Gerrit Jacobus is also only very rarely mentioned in Henriette Hertzsprung–Kapteyn's biography [1]. As we shall see later, he went to study mining engineering in Freiburg; after he had become an engineer, he wandered all around the world. That may well explain why Henriette mentioned him hardly at all in her book, but when David Gill visited in 1900 all of that was still in the future. Gill's special affection for Kapteyn's daughters may be explained by the fact that he and his wife had no children and maybe especially had longed for a daughter, but even then the paucity of mentioning of her brother by Henriette is curious.

In addition to David Gill, two other important astronomers made their appearance in Groningen. Anders Donner was of great importance to Kapteyn's scientific work because of the extensive plate material he produced and sent to Groningen. He did not like traveling, nor did he like writing letters. Kapteyn did not like correspondence either, because the letters he wrote to Donner (those from Donner to Kapteyn have been lost) often begin with extensive apologies for the long period of time that had elapsed since the last letter. But they were certainly good friends and according to Henriette Hertzsprung–Kapteyn's biography [1], Donner stayed several times in Groningen, and probably also in Vries. The first time was in 1896, when Donner had been at one of the progress congresses on the *Carte du Ciel*, and Kapteyn obviously not. Kapteyn sent instructions how Donner could get from Paris to Groningen by train. This lasted about 14 hours; Donner would leave Paris at 8:10 and arrive in Groningen at 22:08 after a transfer in Rotterdam. Despite the great importance of Donner for Kapteyn's work, he is hardly mentioned in other people's writings about Kapteyn. The number of publications of Donner in international journals is small and—with the exception of those with Kapteyn

---

[1] As mentioned before, Gill was Scottish by birth (Aberdeen) and 'lass' for girl is mainly used in Scotland and the north of England.

in the *Publications of the Astronomical Laboratory at Groningen*—generally were no more than routine presentations of observations and simple interpretations. Donner's greatest contribution to astronomy was his efforts for the *Carte du Ciel,* and especially provision of fundamental observation material to Kapteyn.

The third person who has had a crucial influence on Kapteyn's career is Simon Newcomb (Fig. 7.2). Newcomb was born in Canada, but became director of the Nautical Almanac Office of the US Naval Observatory in Washington, established by the American Navy primarily for the support of navigation of ships. The Nautical Almanac Office was responsible for the publication of an annual almanac with positions of the Sun, Moon and stars, and other fundamental astronomical data. This *Ephemeris and Nautical Almanac* was the American version of the *Nautical Almanac of the Royal Greenwich Observatory* in England; the two merged in 1981 to form the *Astronomical Almanac.* Simon

Fig. 7.2 Simon Newcomb (1835–1909). This picture has been taken from Henriette Hertzsprung–Kapteyn's biography [1]

Newcomb was an authority in the field of determining the positions of stars and fundamental constants in astronomy, such as the precession constant (the speed at which the precession of the Earth's pole relative to the stars proceeds) and the Astronomical Unit (the average distance from the Earth to the Sun). In addition, Newcomb was a professor at Johns Hopkins University in Baltimore and author of many popular books on astronomy. My scientific biography of Kapteyn has as subtitle *Born Investigator of the Heavens*, a quote about Kapteyn taken from one of those books by Newcomb. Soon after Kapteyn's work on the *Cape Photographic Durchmusterung* had begun, Newcomb had heard of Kapteyn; when the opportunity arose, he visited the Kapteyns in 1899 in Vries and Groningen. Here too it is instructive to quote a paragraph from Henriette Hertzsprung–Kapteyn's biography [1] (for a picture of Kapteyn at the time see Fig. 7.3):

Simon Newcomb came to Groningen and turned out to be a unexpectedly stiff man, who impressed people by his appearance and by being mostly silent. Kapteyn did not feel very comfortable with him and sometimes got annoyed by it during his visit, no matter how much he appreciated his presence and how instructive and interesting it was for him. Not so Mrs. Kapteyn. She had an easy, natural way of dealing with important people, which was enjoyable and magnificent and at the same time great. She did not allow herself to be impressed by anyone, because she never worried about the impression that she made, but was very careful to make things pleasant and enjoyable for her guests. She immediately liked him; his silence did not bother her, because she had enough to say for herself and mastered the English language very well. 'He is a king among men', she used to say, and in amazement the children saw that he was allowed to stretch out and put his tired, dusty feet on the beautiful golden chairs that were the sanctuary in their mothers' lounge. She baked buckwheat cakes for his breakfast, because he loved that so much, and was the nicest hostess an stern American could think of. The amazement was ours. At that time Newcomb wrote to Gill: 'I have read a letter of you to Mrs. Kapteyn', and apologized by saying; 'It is hard to keep anything from so delightful a woman'. Gill copied this remark back to the Kapteyns, who were enormously pleased with it. Newcomb had a peculiar dry humor, which the children were surprised to find in such an impressive man. He was big and sizable, with heavily wavy white hair, which was eye-catchingly beautiful. He said his hair was his wife's pride. 'Later my name will be mentioned in the lexicons as 'The man who had the most beautiful head of hair. Seems to have been an astronomer'.

Fig. 7.3 Kapteyn in 1908. Centraal Bureau voor de Genealogie, collectie Veenhuizen [69]

7.2   St. Louis

Since the middle of the nineteenth century world exhibitions were held, starting in 1851 in London and 1855 in Paris. These usually had a theme, such as for the first two: 'Industry of all nations' and 'Agriculture, industry and art'. The first one in London is still known from the especially built Crystal Palace. With some regularity new world exhibitions were organized, often in Paris or London and more and more on the occasion of important commemorations. The first American edition took place in 1876 in Philadelphia, on the occasion of the centenary of American independence. For the fourth French World Exhibition was in 1889, one hundred years after the French Revolution, the Eiffel Tower was built. A few years later, the United States wanted to celebrate the centenary of the 'Louisiana Purchase' of 1803 with another world exhibition.

The Louisiana Purchase was the purchase by the United States of an enormous piece of land from France, more than two million square kilometers. It stretched in the form of a wedge, starting around New Orleans in Louisiana on the Caribbean Sea, stretching out to the width of about the current states Montana and North Dakota and even a bit across the current Canadian border; the area comprised fourteen of today's states or parts thereof. The amount involved was 15 million $ (that is, between three and five hundred million Euros or dollars today); the purchase roughly doubled the surface area of the United States. The World Exhibition was finally held in 1904, under the name Louisiana Purchase Exposition, popularly known as the Saint Louis World Fair, after the city where the event took place. Formally, it was the fourteenth World Exhibition, which lasted almost the entire year of 1904. The Summer Olympics of that year were also held there, although participation from Europe was not very large due to the high cost of travel and accommodation; for example, there was no participation from the Netherlands.

As part of the exhibition, congresses were also organized, the largest in the week from September 19 to 24. The theme of this 'International Congress of Arts and Sciences' was the advancement of mankind since the Louisiana Purchase and covered all possible fields of science, technology and art. There would be about one thousand participants. The group of organizers was chaired by Simon Newcomb. On July 15, 1903, about a year before the event, he sent a letter to Kapteyn, announcing that he would be officially invited to give a lecture at the congress. Newcomb offered Kapteyn a sum of five hundred dollars (now that would be just over eleven thousand €) to cover his expenses, and remarked that he would not have to spend all of it if he came by himself. Enchanted as he apparently was by Mrs. Kapteyn since his visit to Groningen and Vries, he suggested that she could come along, although she would have to carry her own expenses.

Traveling was a costly affair at those times, partly because it was time consuming and there were a lot of overnight stays and meals to pay for. After some hesitation, Kapteyn accepted the offer and announced that if he could afford it, he would bring his wife along. Newcomb invited them to visit him and his wife in Washington, which the Kapteyns indeed did on the way back from the Congress.

During the Congress, which was held on the Washington University campus in St. Louis, and for which the facilities were barely sufficient, three astronomy sessions were held at the Astronomy and Earth Science Symposium, each with two lectures (see Fig. 7.4). The most important American astronomers were present and gave a lecture or chaired a session. Jöns Oskar Backlund (1846–1916), director of the Pulkovo Observatorium near St. Petersburg, and Herbert Hall Turner (1868–1938) from Oxford also gave lectures. Important for the further course of Kapteyn's career were Edward Charles Pickering (1846–1919), director of the leading Harvard College Observatory near Boston, and

Fig. 7.4 The logo (left) and the title plate of the publication of the congress in astronomy and earth sciences in St. Louis in 1904. On the right Phoebe is depicted: 'Phoebe—photo-engraving of a painting by Louis Perrey—, in Greek mythology the special name given to Artemis, goddess of the Moon, twin sister of Apollo, who was called Phoebus, the God of the Sun. No other goddess, except Venus, surpassed Phoebe in beauty.' [70]

George Ellery Hale (1868–1938), at that time director of the great Yerkes Observatory near Chicago. Kapteyn's lecture was the second of the morning session of September 21, 1904.

7.3   Star Streams

Kapteyn had kept the title of his presentation general: *Statistical methods in stellar astronomy*. He began with a long digression into the problems of astronomy in understanding the distribution of the stars in space, the absolute necessity of the knowledge of the distances of stars, and the fact that parallaxes, except in a few cases, were too small to measure reliably. But by now it was known that the Sun moved relative to the stars; that made a statistical approach possible. Kapteyn also mentioned the inevitable fundamental assumption that the stars moved randomly through space, with the same average speed everywhere and without a specific preferred direction. And he showed that the available data implied that the number of stars per unit volume decreased as a function of the distance from the Sun. He also showed what the distribution of the velocities of the stars in space was. Next he talked about the two other fundamental assumptions that had to be made, namely that everywhere the stars had to be distributed in the same way over luminosity and that there should be no absorption of light in space. About the latter he said that he did not agree with the results of a study by George Cary Comstock (1855–1934), who at that time was director of the Washburn Observatory in Madison, Wisconsin, and he would soon discuss that in a journal paper. Comstock was one of the listeners; he had chaired the first session of that morning.

Undoubtedly this introduction was an impressive presentation, with which Kapteyn must have reaped great admiration. But then he really started. He now raised the issue of the fundamental assumption—that he had made without any justification—that the spatial velocities of stars were on average the same everywhere and that there would be no preferred direction. I will discuss this using the same figures (here Figs. 7.5 and 7.6 that Kapteyn used in St. Louis in 1904. In Fig. 7.5 he illustrated the fundamental hypothesis in the upper left part (indicated by 'P'). In each direction the average velocity is the same and all velocities are uniformly distributed in terms of direction. But as seen from Earth (or the Sun) it is slightly different, because we move in space relative to all the other stars. Then you get the panel at the bottom left ('Q' in Fig. 7.5). Stars

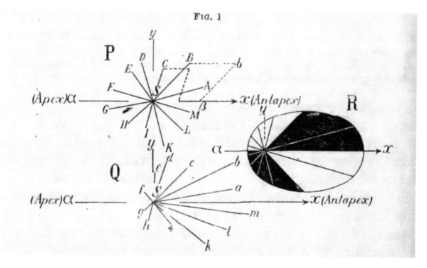

Fig. 7.5 Kapteyn illustrated with this figure in St. Louis in 1904 what the distribution of proper motions over the sky should be if the stars moved randomly and on average equally fast in all directions through space and if the Sun had a velocity relative to the average of all stars. From [70]. Kapteyn Astronomical Institute, University of Groningen

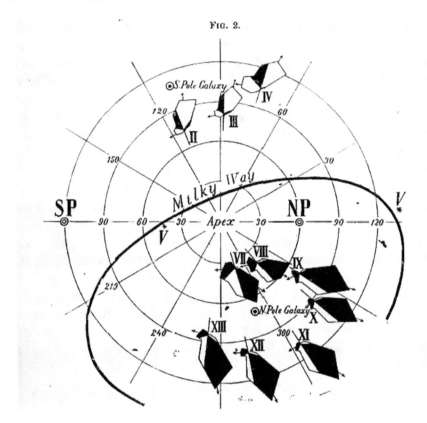

Fig. 2.

Fig. 7.6 With this figure Kapteyn illustrated in St. Louis in 1904 the distribution of proper motions in some areas of the sky of bright stars. This is different than expected according to Fig. 7.5 and resulted in his discovery of the two Star Streams. From [70]. Kapteyn Astronomical Institute, University of Groningen

that move in the same direction as the Sun—in the figure to the left—now have on average a lower velocity relative to us; stars that move in the opposite direction (to the right) have a higher velocity. That pattern should be apparent in every direction on the sky, if you knew the velocities of all stars.

In the panel on the right ('**R**') Kapteyn schematically drew the prediction of the fundamental assumption. If you plot all the proper motions in a certain area on the sky and calculate the mean values for the proper motion, then you should get the pattern **R**. To emphasize the symmetry Kapteyn had blackened two sectors in the figure. Kapteyn now did this for 2400 stars from Bradley's catalog, dividing the sky up into 28 areas. The result can be seen in Fig. 7.6, where for the sake of clarity he showed only ten such areas. This is a projection of the sky on a flat surface with the Apex in the middle. The circles around it lie on the sky respectively 30°, 60°, 90° and 120° from the Apex. He used the notations **NP** and **SP** to indicate where in this representation the celestial poles lie, and the Milky Way with a thick, curved line.

What one now would expect is to see the pattern of Fig. 7.5 **Q** would appear in each of the ten areas, pointing away from the Apex and displaying the corresponding symmetry in the black and light sectors. The patterns of course point away from the Apex (that is how it was found in the first place), but the symmetry is totally missing. The Sun moves through space in the direction Apex-Antapex, but apart from that there is absolutely no sign of a full symmetry in the distribution of the motions of the stars. So Kapteyn showed that there are one or more preferred directions and that was a very remarkable result.

Further analysis by Kapteyn showed what the situation is: the stars move in general terms in two separate groups, which Kapteyn called *Star Streams*. The directions of the two streams were about 125° apart. But if he then corrected those directions of the two streams for the motion of the Sun through space, then they came to lie more or less in the Milky Way and roughly diametrically opposite from each other. In Fig. 7.6 these are indicated with the symbols 'V', one at the far right and the other to the bottom left from the letter **M** for the Milky Way. So the conclusion was that there are two *opposite* streams in the Milky Way.

You need to know the average distance of the stars involved to estimate how fast those Star Streams move. At St. Louis, Kapteyn could not go beyond the rough estimate that that was about 10 to 20 km/s. Later he found that the streaming velocities in space were about 20 km/s, so they had a velocity of some 40 km/s relative to each other.

This discovery of the Star Streams, which had not been announced in advance, must have made an enormous impression at the time. The publication of the lectures during the congress in St. Louis lasted until 1908, but the news went around like wildfire. It has been said that it was the best-known discovery, everyone had heard of well before it appeared in print. Kapteyn's collected papers do not even contain a final reprint of this article[2]—apparently he did not have one in his possession, but there are at least three copies in Groningen as stencils on thin, translucent paper, with the figures inscribed with a pen. Two of them are part of the volumes with articles in which Kapteyn had collected his papers (see Fig. 0.3), and one is in the archives of his successor Pieter van Rhijn. It is a fact that in 1905 Kapteyn also presented the concept of Star Streams at a congress in Cape Town, but from that only a, not very detailed, summary appeared in print.

What certainly helped an almost immediate acceptance in a broad circle of the concept of Star Streams, was that independent confirmation soon came.

---

[2]It is available electronically via [70], beginning on page 396.

With data for a different set of stars, British astronomer Arthur Stanley Eddington (1882–1944) found that Kapteyn's conclusions were correct (Fig. 7.7). Where Kapteyn used bright stars over the entire sky, the data available to Eddington concerned fainter stars over smaller areas of sky. Eddington had just obtained his PhD at the University of Cambridge and would become one of the most influential theoretical astronomers of the twentieth century.

However, quickly another interpretation was given, which would only convincingly turn out to be the right one much later on. The German astronomer Karl Schwarzschild (1873–1916), director of the Sternwarte in Göttingen (Fig. 7.8), suggested that these were not two physically separate, opposite streams, but that this apparent streaming effect was in fact caused by an asym-

Fig. 7.7 Arthur Stanley Eddington (1882–1944). Eddington is probably best known for his theories about the structure of stars. This picture comes from the *Album Amicorum*, presented to H.G. van de Sande Bakhuyzen on the occasion of his retirement as professor of astronomy and director of the Sterrewacht Leiden in 1908 [47]

Fig. 7.8 Karl Schwarzschild. He has become particularly famous for his work on the radius of a black hole, which is named after him [71]

metry in the random velocities of the stars. All stars moved through space in all directions, but in the direction of the apparent streams the average random velocity of the stars would be larger than in other directions. In reality, the stars go around the center of the Milky Way at a velocity of about 220 km/s in the vicinity of the Sun, but the random velocities are much smaller relative to each other. These velocities are larger in the direction to and from the center of the Galaxy than in the direction of rotation, and they are even smaller perpendicular to the plane of the Milky Way. Eddington found, in accordance with this, that the number of stars in the two Streams of Kapteyn was about the same. But both he and Kapteyn remained preferring the idea of two streams because of the observation—which later turned out to be incorrect—that the types of stars in the two streams were very different. In Schwarzschild's model this would be difficult to explain. Only after Kapteyn's death did Eddington become convinced of the correctness of Schwarzschild's idea; Kapteyn always continued to believe in his Star Streams.

By the way, Schwarzschild is best known for his derivation of the 'radius' of a black hole from Einstein's theory of General Relativity of 1915. He published his calculations just before his death in 1916. He is also the father of Martin Schwarzschild (1912–1997), who worked mainly in Princeton and formulated, among other things, the theory of the structure and evolution of stars during their later stages.

7.4   Further Travels in America

Kapteyn was one of the first Dutch scientists, certainly the first astronomer, to attend scientific meetings in the United States, at least an astronomer with such a prominent status. By the way, he was not the only Dutchman who was invited to the congress in St. Louis. The other one was the plant physiologist and geneticist Hugo de Vries (1848–1935) of the University of Amsterdam. Dutch science was still strongly focused on Germany and the United Kingdom; Dutch astronomers published as mentioned above mainly in the *Astronomische Nachrichten* of the (German) Astronomische Gesellschaft, and usually in German. It was around this time that Kapteyn strongly developed an Anglo-Saxon orientation. He was aware that Amerindian astronomy was booming, and from 1902 onward he published more and more in American journals.[3] Of course, the *CPD* was already in English at the time, but more remarkable is that the *Publications of the Astronomical Laboratory at Groningen* were in English, while the *Annals of the Sterrewacht Leiden* continued to be published in German until Willem de Sitter became a professor there.

Kapteyn was involved in a second meeting while he was in St. Louis. George Hale was building a large solar observatory on Mount Wilson near Los Angeles, and organized a meeting in St. Louis to coordinate solar research. Knowing that Kapteyn had already been invited to St. Louis by Simon Newcomb, Hale wrote to Kapteyn whether he would like to attend that second meeting as a representative of the Netherlands Academy of Sciences. He also invited Kapteyn to visit the Yerkes Observatory at Williams Bay near Chicago during his trip to the United States. George Hale was the director there, but would be in California at the time; deputy director Edwin Brant Frost (1866–1935) would be host to Kapteyn. It turned out later that Kapteyn had asked Hale, who had built the largest refractor telescope in the world at Yerkes—i.e. with a lens as the objective (40 in. or about 1 m in diameter) and not a mirror—about the optimal focal length to measure parallaxes and proper motions. The letters from Kapteyn and Hale crossed. The Kapteyns indeed visited Williams Bay.

---

[3] Kapteyn's last article in the *Astronomische Nachrichten* dates from 1910, but that was written in English.

In August 1904 the Kapteyns left home. In Henriette Hertzsprung–Kapteyn's biography [1], their departure from the Netherlands is described as follows:

> Mrs. Kapteyn accompanied him. Characteristic of these two simple people was their departure. They left late in August, together by bicycle from their second home in Vries, to Groningen to take the express train to Rotterdam. The children were sitting on the fence and waved them goodbye. So began their first great journey to America, Kapteyn to conquer the astronomical world, Mrs. Kapteyn, to make the many distant friends in whom she would find like-minded friends. Arriving in St. Louis, they were received by the Dutch Consul, who quartered them in primitive wooden, but not very dignified quarters, since everything was overcrowded. They had arrived just in time to attend the opening of the astronomical congress, and had quietly settled in the back of the room. Newcomb, the president, had discovered them and immediately came to Kapteyn to lead him to the board table, where the big shot astronomers were gathered. He was greeted with great cordiality and respect by all, and felt immediately taken in as a welcome and honored guest.

After the Congress in St. Louis, the Kapteyns paid a visit to the Newcombs in Washington. There a deep friendship developed between the ladies Kapteyn and Newcomb. Mrs. Newcomb, Mary Caroline Hassler (1840–1921), and Elise Kapteyn corresponded regularly with each other from that time on (often as a supplement to letters from their husbands about astronomical matters); some of those letters have been preserved in the archives of Newcomb. On October 31, 1904, not long after her return, Elise wrote Mrs. Newcomb a letter with personal details. She expressed her thanks to them and to the many friends they had made, but went on to mention that their daughter Henriette had broken off her courtship and that it had caused them much grief. On the boat they had met a teacher from Chicago, who was going to Bologna via Paris; she had in the end come along with the Kapteyns. Through her advice Henriette was persuaded to drop out of law school and study English and French literature. Elise also announced that the eldest daughter's studies were going well and that their son was about to go to Germany to study mining engineering. The eldest daughter, who would be named Noordenbos after her marriage, had already claimed a butter tub with the letter 'N' engraved on it, which the Newcombs had donated.

During their visit to Washington, the Kapteyns were received in the White House by the president, Theodore Roosevelt (1858–1919). Henriette Hertzsprung–Kapteyn [1] wrote about this:

On their return journey they visited Newcomb in Washington, where they would also attend the big reception at the White House. They rode in Newcomb's equipage with two horses and black palfrenier, and were introduced to president Roosevelt at the same time as Hugo de Vries, who was also in Washington then. He had an appropriate word and a handshake for each of them: 'Ah Mr. de Vries – I suppose we are cousins, because the name de Vries is in my family.' This, of course, with the necessary pride, because Dutch names meant status in America. For Kapteyn he had a friendly remark about astronomy, and also this remarkable moment belonged to the past.

7.5   Parallaxes Again

The discovery of the Star Streams was a serious problem for the program that Kapteyn had devised to determine the spatial distribution of the stars. Of course he did not have to give it up completely, but it all became much more complicated. The next steps were the following. In the first place there had to be counts to much fainter magnitudes than previously available. Furthermore, measurements of as many parallaxes and proper motions as possible, and of course a better determination of both the position of the Apex of the Sun and its spatial velocity, and in addition now also the direction and velocities of the Star Streams. And it became more and more clear that stars among themselves had very different properties, which were not yet understood, but this had to be determined.

In *Publications of the Astronomical Laboratory at Groningen* number 8, Kapteyn had determined as well as was possible the average parallax of stars as a function of apparent magnitude and proper motion. With this statistical estimate of stellar distances he was able to make a rough estimation of the distribution of stars in space and that resulted in 1901 in a first estimation of the run of star density with distance from the Sun. During a lecture for the Netherlands Academy of Sciences, he presented it for the first time. First he determined the 'luminosity curve', which he had assumed was the same everywhere, and using that the density of stars. What became clear was that the stellar density decreased as one moved away from the Sun. It was known that stars had very different spectra and could therefore differ significantly from one another in luminosity. In 1901 this had not yet been properly determined, but Kapteyn divided the stars into two main groups and did a first, very preliminary investigation into how the laws of brightness and density depended on this. He published this first attempt a year later in *Publications of the Astro-*

nomical Laboratory at Groningen 11; this research was also briefly presented in St. Louis.

However, it was clear that there was not enough information to take the next step. Kapteyn began an extensive set of investigations into the distances of stars, together with his assistants Willem de Sitter and his second PhD student Herman Albertus Weersma (1877–1961) (see Fig. 7.9). He obtained his PhD in 1908; although journalist and amateur astronomer Cornelis Easton had obtained a PhD in 1903, Herman Weersma was only Kapteyn's second student, since Easton's degree was awarded *honoris causa*. Weersma's work was largely based on plates by Anders Donner. Among other things he tried to measure parallaxes in star clusters, such as the double cluster h and χ Persei with Anders Donner and Willem de Sitter and the nearby cluster Hyades with Donner. In these and other fields, measurements were also made and studies conducted on individual stars.

Fig. 7.9 Herman Albertus Weersma. This picture comes from the Album Amicorum, presented to H.G. van de Sande Bakhuyzen on the occasion of his retirement as professor of astronomy and director of the Sterrewacht Leiden in 1908 [47]

In the case of the double star cluster h and χ Persei (see Fig. 7.10), the brightest stars can just be seen with the naked eye; the clusters were already known in ancient times. They are at a distance of about 2.3 kpc (7500 lightyears) from us and of the order of 10 to 15 million years old. So they are young by astronomical standards. The distance as we know it now comes down to a parallax of 0″.0004, which Kapteyn, Donner and de Sitter were far from able to measure. They estimated their errors for parallax measurements of individual stars as 0″.02, or 50 times the value required for the double cluster. The distance had to be found from other methods.

The Hyades constitutes the star cluster closest to us; the cluster is a little over 600 million years old. The brightest stars can be seen with the naked eye, but because the cluster is so close (some 45 pc or a parallax of 0″.022) it is not easy to recognize the cluster as such. In the sky, it spans more than the constellation Taurus, the Bull. In Fig. 7.11 we see an area of about 20° × 30°. Below the center we see the 'tuning fork' formed by the brightest stars of the constellation Taurus. The brightest star, Aldebaran, the 'red eye of the Bull', does not belong to the Hyades. However, most of the other brightest stars in the photograph are part of the Hyades. At the top right of Fig. 7.11 we see the cluster Pleiades or the Seven Sisters. It does not belong to the Hyades, and is at a larger distance (about 140 pc) and much younger (of the order of 100 million years) than the Hyades.

For the Hyades, proper motions can be used to determine which stars belong to the cluster and which do not. These proper motions should be more or less parallel for cluster members. Herman Weersma was involved in this by collecting proper motions from the literature at an early stage. A major study

Fig. 7.10 The double star cluster *h* en χ Persei. National Optical Astronomy Observatory [72]

Fig. 7.11 The Hyades cluster encompasses the largest part of the the constellation Taurus. The brightest star is Aldebaran (α Tauri), the red eye of the Bull, but that star is not part of the Hyades. On the top right the cluster Pleiades, of which the brightest five or six stars are visible to the naked eye. Hubble European Space Agency Information Center [73]

was performed using plates taken by Anders Donner and was carried out by Willem de Sitter. In addition, plates from the Bonner Sternwarte were used, taken by Karl Friedrich Küstner (1856–1936). This resulted eventually in a large number of individual parallaxes, but because the stars from the cluster had to be at the same distance, it was possible to determine an average parallax. The answers were encouragingly similar, +0''.024±0''.010 for the Helsingfors plates from Donner and +0''.023±0''.0025 for Küstner's plates from Bonn. The best current value is 0''.0215±0''.0028. The results were published in 1904 in issue 14 of the *Publications of the Astronomical Laboratory at Groningen.*

Not much later, in 1908, a study by the American astronomer Lewis Boss (1846–1912) of Dudley Observatory in Schenectady, New York was published [74]. He used his own proper motions with a method which is in fact the same as that used to determine the Apex of the Sun and which lies at the basis of the secular parallaxes proposed by Kapteyn. This method can be traced back to Auguste Bravais. Now Boss had found that the Hyades cluster extends over a much larger area in the sky than Kapteyn was aware of, and that enabled him to go one step further. Figure 7.12, taken from Boss' publication, shows that if you extend the proper motions, they come together in almost one single point, about 25° from the cluster itself. This is the same as the Apex of the Sun, except that what we see here is not the direction of the Sun's motion relative to the average of the stars in the neighborhood, but relative to the Hyades.

This opens up new possibilities. The angle on the sky of a star in the Hyades to this Apex corresponds to the angle at which the spatial velocity of that star is projected onto the line of sight. If you now measure that star's velocity along this line (the radial velocity), we can calculate what its total spatial velocity is, but also—and that is the important point—what the velocity is perpendicular to the line of sight (tangential velocity). Now, you know how much the proper motion on the sky is, from that and the tangential velocity follows immediately

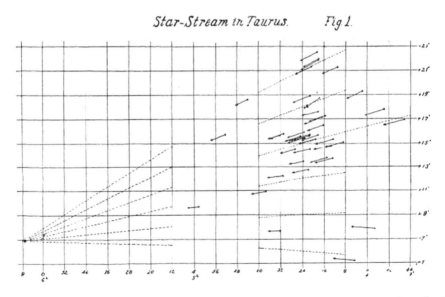

Star-Stream in Taurus.   Fig 1.

Fig. 7.12 Proper motions in the Hyades according to Lewis Boss in 1908. The length of the arrows corresponds to a proper motion of over 50,000 years [74]

the distance. Karl Friedrich Küstner had previously measured the radial velocity of three stars in the Hyades (in these cases almost exactly 40 km/s), and from Boss' determination of the Apex for the Hyades followed a spatial velocity of the Hyades relative to us of 45 km/s and a parallax of $0''.0253$. This method of distance determination for a nearby cluster is called the 'moving cluster method'. The Hyades today are still vital for the distance scale in the Universe, because as the nearest cluster it is a reference point for more distant star clusters.

But there was more work to be done. Kapteyn himself did extensive research in order to improve the accuracy of the precession constant. As we have seen, the coordinates of stars change as a result of precession; the spinning motion of the rotation axis of the Earth. As a result the vernal equinox (both equinoxes of course) moves along the equator and it was absolutely necessary to know how fast this change occurred (the precession constant) in order to be able to measure proper motions, which are after all based on changing positions on the sky. In addition, in order to be able to use secular parallaxes, it was necessary to split the proper motions of stars into the component in the direction Apex-Antapex, and the one perpendicular to it. Kapteyn did not like quick and dirty work and, if necessary, was prepared to take on enormous amounts of calculation, which was done by assistants (calculators). For the Bradley-stars he calculated those components for six different combinations of the precession constant and the position of the Apex. Weersma completed his PhD thesis in 1908 as part of this project. It was entitled *A determination of the Apex of the solar motion according to the method of Bravais*.

Weersma did more important work in support of Kapteyn's program, but left astronomy in 1912. He was very much interested in philosophy and in the long run astronomy was no longer interesting enough for him. In order to have more time to better follow his interests, he became a mathematics teacher. He converted himself to socialism and wrote books on Marxism and other philosophical themes involving dialectics and logic.

All this work of his PhD students and assistants was part of the preparation for Kapteyn's eventual attempt to solve the problem of the construction of heavens. But first of all, what was needed was a reliable and extensive inventory of the stars in the sky, including their properties. Kapteyn had thought about this at length before he went to St. Louis.

7.6   Plan of Selected Areas

Not long after the turn of the century Kapteyn had already realized that it was necessary to systematically determine the distribution and properties of stars across the entire sky. But to significantly fainter magnitudes than available in the existing star catalogs. He defined for this his *Plan of Selected Areas*. In essence, he selected a limited number of small areas all over the entire sky, which would allow a representative census of the stars to be determined. In the end it resulted in 206 uniformly distributed areas. To produce an inventory of stars in these areas was a gigantic undertaking and in order to complete it successfully, he needed the cooperation of a large number of observatories, especially those with the best telescopes and instruments. It takes tact (and time) to accomplish such a thing.

Kapteyn began discussing his plan with many astronomers. His visit to St. Louis also played an important role in that respect; the directors of the most important American observatories—as well as some others—were there and Kapteyn must have talked to them about his plans. After returning to Groningen, he wrote a draft of the Plan, which circulated for several years among astronomers and directors of observatories. The matter must also have been the subject of much deliberation when a year after St. Louis, in 1905, he visited South Africa—the only trip he would ever make to the southern hemisphere in his lifetime. It had been Kapteyn's old wish to visit David Gill (see Fig. 7.13) in his own surroundings and to see the southern starry sky, which he had after all mapped out with him. The occasion was a meeting of the *British Association for the Advancement of Science*, that would meet in Cape Town and Kapteyn had been invited to that meeting. This organization, now called the *British Association*, was founded in 1831 with the aim of promoting knowledge about science and to foster the interaction between scientists. The Association met once a year. I quote again from Henriette Hertzsprung–Kapteyn's biography [1]:

> In the summer of that year Kapteyn traveled to South Africa, at the invitation of the British Association, which held its large international meeting there. De Sitter, who was then his assistant, accompanied him. Several other astronomers traveled also along with him on the same ship: Backlund, director of the Pulkovo observatorium (Russia), Donner, Hinks and Cookson, the latter both English astronomers. They talked a lot during the trip and founded 'The Astronomical Society of the Atlantic' for the duration of the sea voyage. The main topic of their discussions probably was the *Plan of Selected Areas*, of which each of them had brought a copy of the draft. Hinks made notes of what was proposed, and Kapteyn could be satisfied with the results of the discussions. Everyone wanted their part in the work. In South Africa he was able to win even more participants for his plans.

> For a long time it had been his wish to see Gill in his own observatory, because one only gets to know a man completely when one meets him in his working place. With pride he was shown the observatory, which Gill had brought so much

Fig. 7.13 David Gill. This picture comes from the *Album Amicorum*, presented to H.G. van de Sande Bakhuyzen on the occasion of his retirement as professor of astronomy and director of the Sterrewacht Leiden in 1908 [47]

to flourish, and he saw his friend as the great man, as president of the British Association and host.

Kapteyn finally published his Plan in 1906, after he had received pledges of cooperation from the most important persons. Figure 7.14 shows the distribution of the 206 *Selected Areas* across the sky. This seems more regular than it in fact is; the exact locations were chosen to avoid bright stars and other extended objects such as nebulae and the like. The publication specified the data that Kapteyn proposed to be measured. In addition to accurate magnitudes for 200,000 stars in two wavelength regions (so that there would also be information about the colors), these were the parallaxes and proper motions for 20,000 of these stars, as well as spectra, so that the type of star could be determined, and for as many stars as possible radial velocities. The dimensions

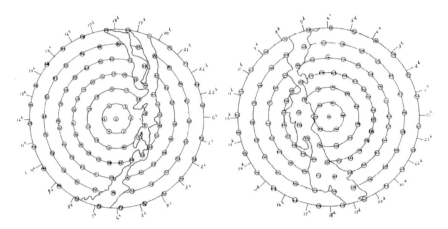

Fig. 7.14 The distribution of the *Selected Areas* over the sky. On the left a representation of Areas 1 through 115 in the northern sky with the pole in the center, the circular edge is the equator. On the right, Areas 92 through 206 in the southern hemisphere, so that the two outer rings are the same. The irregular lines outline the Milky Way. Kapteyn Astronomical Institute, University of Groningen

of the areas were not strictly defined; in the Milky Way, where the stars are close together in the sky, they would be smaller than outside the Milky Way. Kapteyn added the total amount of background light received from the sky in various areas. If one could measure this—and I will come back to it later—and correct for contributions other than the total light from the stars, you would have an important boundary condition for construction of a model for the stellar distribution in space.

Letters of support and pledges of participation came from all over the world: Edwin Brant Frost, now director of the Yerkes Observatory, who had received the Kapteyns as a Deputy after their visit to St. Louis; Karl Friedrich Küstner, director of the Bonner Sternwarte; George Cary Comstock of the Washburn Observatory in Madison, Wisconsin. But prominent and of fundamental importance to the success of the Plan were especially Edward Charles Pickering, who was the founder of the observatory in Madison, Wisconsin, now director of the Harvard College Observatory, and George Ellery Hale (Figure 9.5), now director of the Mount Wilson Observatory near Pasadena, California. Both men deserve a separate description.

Edward Pickering (see Fig. 7.15) had studied at Harvard University and became director of the Harvard College Observatory in 1877. He was a pioneer in obtaining and investigating the spectra of stars. The physician and amateur astronomer Henry Draper (1837–1882), who also lived in that area, was one of the first to experiment with photography in astronomy and is described

Fig. 7.15 Edward Charles Pickering (1846–1919), director of the Harvard College Observatory [75]

to actually use it in his private observatory where he managed to obtain the spectrum of a star, namely the bright star Vega. This is α Lyrae, the brightest star in the constellation Lyra and fourth brightest star in the sky. He noticed that there were lines in the spectrum that were reminiscent of the spectrum of the Sun. Hot, young stars for example have spectra that are dominated by lines of helium and hydrogen, but in stars like the Sun these are less prominent and the spectrum is dominated by atoms and ions (ions are atoms that have lost one or more electrons) of other chemical elements such as sodium, calcium, iron, and so on. The systematics of the stellar spectra has been extensively researched under Edward Pickering, who thereby turned Harvard College Observatory into a prominent observatory.

Draper recorded many more spectra of stars; after his death, his widow set up a fund under the name of Henry Draper Memorial to make it possible to further study this matter. This resulted in the *Draper Catalog of Stellar Spectra*, published in 1890, with classifications of more than a thousand stars. This classification was done in a newly conceived system and was mainly carried out by Williamina Paton Stevens Fleming (1857–1911). It was a preliminary study and Edward Pickering and his female assistants, often referred to as 'Pickerings harem', among which Antonia Caetana de Paiva Pereira Maury (1866–1952) and Annie Jump Cannon (1863–1941), completed the catalog and in the course of that defined in an elaborate classification system, the Harvard classifications . The *Henry Draper Catalog* and supplements, published between 1918 and 1936, are still in use today.

Edward Pickering had also given a presentation during the congress in St. Louis and Kapteyn undoubtedly spoke extensively with him and made him enthusiastic about his Plan. The Harvard College Observatory also operated a southern observatory, the Boyden Station in Peru, first located near Lima, later inland near Arequipa. It was built with money from engineer and inventor Uriah Atherton Boyden (1804–1879). Edward Pickering finally pledged his cooperation to Kapteyn and agreed that all *Selected Areas* would be photographed with his telescopes on both hemispheres.

But Kapteyn had to pay a price. Edward Pickering wanted something to be added to the Plan, namely that there would be more Select Area's spread across the Milky Way, especially in areas where star-density was high or where there were specific nebulae. That would provide more information about the Milky Way, and provide better sampling. Kapteyn resisted this addition because of the extra work and time it would take, but eventually he had to agree to keep Pickering on board. Pickering then set to work energetically, and all plates had been recorded in 1914. Kapteyn and his people set about measuring those plates. Eventually the results of this *Harvard-Groningen Durchmusterung* were published in three volumes in the *Annals of the Harvard College Observatory*, in 1918, 1923 and again 1923, the last two as far as Kapteyn was concerned posthumously. The results of the *Special Plan* were published only in 1952, under the direction of Kapteyn's successor Pieter van Rhijn.

George Hale had founded the Yerkes Observatory near Chicago in 1897 with money from the broker and financier Charles Tyson Yerkes (1837–1905). Hale soon built the largest refractor telescope in the world. A refractor telescope has a lens at the top of the tube as the primary optical element, in contrast to a reflector, which has a mirror at the bottom. A lens has the disadvantage, unlike a mirror, that it can only be supported at the edges. This means that it deforms when the orientation of the telescope changes. Furthermore, the light must pass through the glass of the lens, partially absorbing it. A mirror reflects, and although some of the light is always lost, that loss is minimal if the mirror is kept clean. In modern telescopes, a thin layer of aluminum (in some cases silver) is applied to the mirror by atomization in a vacuum chamber. Breaking of light through a lens, as opposed to reflection, also depends on the wavelength. Blue light is refracted more than red light, so that with a simple lens the image is not formed in exactly the same place in different colors. This is called chromatic aberration. This can be solved with achromatic lenses, which consist of several parts stuck together, that are made of various materials with different refractive properties. With mirrors there is spherical aberration because, with a spherical surface of the mirror, the image is formed in a different place when it is reflected at the edge than when it is reflected in the center. This can be solved by making the surface parabolic, but that is difficult and also makes the field of view, the area over which good stellar images are formed, smaller.

The Yerkes telescope had a primary lens with a diameter of 102 cm (40 in.). George Hale was originally a solar astronomer, but in order to understand the Sun, he felt he had to study other stars as well. In 1903 he was in California, where he visited among others Mount Wilson near Pasadena as a potential location for an observatory. Edward Pickering had also had had an eye on that spot earlier. George Hale secured the support of the Carnegie Institution of Washington (presently 'of Science'), founded in 1902 by steel magnate and philanthropist Andrew Carnegie (1835–1919) to promote science. On the land that George Hale was able to purchase with the funds the Institution provided, he initially installed a few solar telescopes from Yerkes and then had two enormous solar towers built; since a long focal length is required for the this, those telescopes were built vertically, with the focal point below ground, where it is cooler. But George Hale also wanted to do night astronomy on Mount Wilson; he made plans for a telescope with a mirror diameter of no less than 60 in. (152 cm). That would then be the largest in the world. The conditions for astronomical observations on Mount Wilson were superior to those at other places, due to the usually clear weather in Southern California and the calm atmosphere at an altitude of 1740 ms. George Hale, as we will see, proposed to use his 'giant telescope' of 60 in. aperture for the *Plan of Selected Areas* of Kapteyn.

Fig. 8.0 This picture was used by Cornelius Easton as an illustration in his article *Personal Memories of J.C. Kapteyn* in the Dutch amateur periodical *Hemel and Dampkring*.

8

In the Meantime in Groningen

We try in vain to describe a man's character,
however, let his acts be collected
and an idea of his character will be presented to us.
Johann Wolfgang von Goethe (1749–1832)

The mediocre teacher tells. The good teacher explains.
The superior teacher demonstrates. The great teacher inspires.
William Arthur Ward (1921–1994)

8.1   Lectures

In Kapteyn's time, even more than today, teaching was central to the work of a professor. Research actually came in second place. This also applied to Kapteyn, and he spent a significant part of his time on lectures and other education. Kapteyn must have belonged to the last category of teachers from William Ward's quote above, as witnessed by the statements of his pupils and students. Kapteyn paid tribute to his descent from a lineage of teachers.

William Ward was an American author.

The *Groningsche Studentenalmanak* for 1883 mentions:

> From Prof. Kapteyn we learned to have a look at the depths of the universe;
> astronomy always captivates the layman, but one may be convinced that after
> his fascinating lectures, even if it would concern a less popular subject, we would
> still practice this science with zeal and pleasure.

And in the commemorative book on the occasion of the 350th anniversary of
the University of Groningen in 1964, professor of physics Wiepko Gerardus
Perdok (1914–2005) wrote [76]:

> For him, understanding was worth more than knowledge of facts and he preferred
> to show at his lectures how today's well-established certainties used to be problems
> and how they had been solved by his predecessors. All of Kapteyn's disciples
> testified to the rare clarity of his discourse and experienced the inspiring influence
> that emanated from him. The recollection of the way in which they saw the
> esteemed teacher construct a circle will not easily be forgotten: Kapteyn took
> the big compass under his arm at some distance from the blackboard, stormed
> towards it with the sharp point ahead and drove it deep into the wood, saying:
> 'I need a fixed point'!

The later famous astronomer Jan Hendrik Oort (Fig. 8.1) studied physics
and astronomy in Groningen and was so fascinated by the captivating lectures
of Kapteyn that he decided definitively for astronomy. Oort began his studies
in 1917, four years before Kapteyn's retirement. Eventually Oort received his
Ph.D. in Groningen under Pieter van Rhijn. In his inaugural lecture with
which he accepted his appointment to a professorship in Leiden he referred to
Kapteyn as 'my inspiring teacher'. Henriette Hertzsprung–Kapteyn's biography
[1] quotes a part from a written statement by Oort about his inspiring style of
lecturing:

> I have never met anyone who could do this to such an extent, and I believe this
> is the most beautiful way of scientific research. Being in his company, during
> lectures or a colloquium, [a scientific lecture at an institute] with something
> unusually stimulating, most of the time I left with more beautiful and hap-
> pier thoughts than when I arrived. This is because he saw the beauty of nature
> and of science, the interrelationship of which he saw more clearly than anyone
> else... Whereas other lectures would tend to make a first-year student lose his
> confidence, the astronomy course restored and enhanced that.

And in an autobiographical article Oort wrote in 1981 [77]:

Fig. 8.1 Jan Hendrik Oort and his at that time fiancée Johanna Maria (Mieke) Graadt van Roggen at the dinner after defending his dissertation and obtaining his Ph.D. degree in Groningen on May 1, 1926. They married on May 24, 1927. From the Oort Archives

When in 1917, at the age of 17, I began my studies in Groningen I became almost immediately inspired by Kapteyn's lectures on elementary astronomy. Although I had been strongly interested in astronomy since my high school years, and this influenced my choice of the University of Groningen because Kapteyn was there, I was in 1917, still undecided between physics and astronomy as my major direction; I remember that I was so impressed by the way he taught elementary celestial mechanics that I tried to convey my new insights to friends who had likewise entered the University, but were studying humanities. But I do not believe that I succeeded in conferring to them a full appreciation of the fascination of celestial mechanics.

Perhaps the most significant thing I learned –mainly, I believe from Kapteyn's discussion of Kepler's method of studying nature– was to tie interpretation directly to observations, and to be extremely wary of hypotheses and speculations. In the first part of his course, Kapteyn refrained, for instance, from introducing the notion of 'force' to replace the measurable quantity 'acceleration'. His disliked intricate mathematical formulations which prevented one from 'seeing through' a theory; he feared the danger that the formulae might make one lose sight of the essentials. This was, of course, before quantum mechanics brought home the fact that one's insight is insufficiently developed to 'look through' the deeper domains of physical science without the aid of mathematics.

8.2   Anecdotes

Kapteyn was unconventional and progressive in some respects. For example, all his children received a similar education. In the words of his daughter Henriette Hertzsprung–Kapteyn [1]. 'The girls attended the Hoogere Burgerschool [HBS, in the meantime no longer only accessible] for boys and the Gymnasium, which was still an exception at the time. Kapteyn did not think about making any difference between his daughters and his son, who also went to the HBS. For all of them the same choices, everyone the same rights, that is how it was and remained for him.' But, she also mentioned, he was also averse to 'ambition and fighting to be preferred over others.' When his eldest daughter came to tell him, radiantly, that she was No. 1 of her class for which she had worked hard, his answer was: 'Never do that to me again.' With this he showed himself to have a wise and deep insight into the true values of life.'

After high school, all three children went to university. The eldest daughter studied medicine in Groningen, the younger law, for a while in Groningen, but she later switched to English in Amsterdam. The son went to Freiburg to study mining engineering. The fact that he did not go to Delft for this had to do with the lower costs of studying in Freiburg. Professors did earn a decent salary, but many of them, including Kapteyn, did extra teaching in order to be able to finance the costs of their children's studies. The fact that the daughters went to university was still quite unusual; the first female student at a university (that of Groningen) had been Aletta Henriëtte Jacobs (1854–1929), who studied medicine in 1871, hardly a generation before the Kapteyn daughters entered university.

The fact that Kapteyn was an unconventional man in a number of ways may have been a reaction to his strictly religious upbringing. But he did not allow reform at all fronts. His marriage was along conventional lines; his wife took care of the household, while Kapteyn saw the finances as his exclusive responsibility. Kapteyn was also conservative in the formal way he signed letters. David Gill always ended his letters with a personal remark, such as 'Your ever sincere friend, David Gill'. Kapteyn, on the other hand, signed his letters to Gill with something like '(very) truly/sincerely/faithfully yours, J.C. Kapteyn'. Also in letters to other colleagues, such as Newcomb, he signed with initials, while Newcomb wrote 'Simon'.

Fig. 8.2 Cornelis Easton (1864–1929). This picture comes from the *Album Amicorum*, presented to H.G. van de Sande Bakhuyzen on the occasion of his retirement as professor of astronomy and director of the Sterrewacht Leiden in 1908 [47]

Anecdotes about Kapteyn can be found in various articles and other writings that illustrate aspects of his personality. Cornelis Easton (1864–1929) (see Fig. 8.2) was actually a journalist, but also a successful amateur astronomer. In that capacity he even published articles in professional astronomy journals, such as the American *Astrophysical Journal,* on subjects such as the spiral structure of our Galaxy, which he derived from the structure of the Milky Way in the sky. Because of his merits, Easton received an honorary doctorate in Groningen in 1903, with Kapteyn acting as an honorary professor. Kapteyn and Easton were good friends and shortly after Kapteyn's death Easton wrote an extensive article about Kapteyn in the magazine of the association of amateur astronomers and meteorologists, *Hemel and Dampkring* entitled *Personal memories of J.C. Kapteyn.* I quote:

Sometimes he talked about his childhood [...] and that he himself had once sat down on a hanging scaffolding for house painters, as a result of which the thing suddenly started sinking with jerks - how he had gotten off, he did not remember himself anymore. Or how he and one of his brothers, in the period before the Rover safety's, triumphantly drove into his parents' Barneveld; however, no-one had never seen such a bicycle, the population all gathered, so that their way was blocked and the two 'cyclists' soon fell 'like ripe pears' from their high two-wheeled vehicles ....'

The 'Rover safety' was a forerunner of our current bike, with a quadrangular frame and a rear wheel driven by pedals and a chain. Former bicycles had a large front wheel that was pedalled directly on the axle.

The symposium proceedings *The Legacy of J.C. Kapteyn* by myself and Klaas van Berkel [4] contains a chapter by historian Wessel Krul about Kapteyn's personality, in which various anecdotes and character descriptions can be found. The following paragraphs are taken from this.

The dominant trait in Kapteyn's personality seems to have been a sharp sense of economy. Not that he was mean or a miser, far from it, but he obviously hated unnecessary expenses, especially when they only served to underline one's social status. He lived without luxury, he dressed carelessly, and out of principle he always traveled second class.

One evening, as Kapteyn returned home from his laboratory, an apparently inexperienced policeman noticed something unusual about him. He did not seem to know what he was doing, and the disorderly way he was dressed aroused suspicion. Was he drunk? A madman? Or a criminal? Kapteyn soon became aware that he was being followed, and decided to play the game. Instead of walking directly to his house, he took his persecutor on a tour of the town, leading him around all sorts of alleys and byways. Great was the disappointment of the loyal servant of the law, when finally the suspect took his key, bade him farewell and disappeared behind a front door bearing the name of a well-known university professor.

Kapteyn often forgot to take his things from the train. Easton once waited for him at the Maas railway station in Rotterdam, where he lived.

He looked extra plain. His straw hat had lost its original freshness a long time ago; no umbrella in spite of the rain – 'I am not allowed a new one any more', he jokingly said, 'I loose them anyway' - and we shuffled together, not unnoticed, by the crowd of people to my home.

Kapteyn was also a universally recognised and loved figure among the inhabitants of the city of Groningen. The story goes, according to Perdok in the above-mentioned commemorative book on the occasion of the 350th anniversary of the University of Groningen, [76] that on the return journey from a meeting in 'Holland' he forgot his bag when leaving the train at the Groningen station. A salesman chased him to his property with the words (in the local slang): 'Prefesser, je vergeten joen tazze met monsters' (Professor, you forgot your bag with samples). Kapteyn, pleasantly surprised, replied: 'Thank you, but there are no samples in it, you know', to which the helpful travel companion corrected himself: 'Well, joen tazze with steerns dan' (well, your bag with stars then)!

8.3   Character

Henriette Hertzsprung–Kapteyn devoted several pages in her biography [1] to Kapteyn's character, habits and daily schedule. I might paraphrase, but in order to preserve their originality, I prefer to quote directly (in my translation):

> The Groningers would see Kapteyn in the street around 9 o'clock in the morning, on his way to his laboratory, everyone knew him with his characteristic boyish little shuffling steps, which he called the 'heather walk', and which he attributed to his many walks as a youth through the heather. From nine to twelve no one was allowed to disturb him and in those three daily hours with concentrated work at his writing desk, and his mind would take a great flight as the major problems progressed. He had great concentration. Nothing could distract him. At home he worked best in the living room, where he felt the warm coziness as a stimulus. 'Concentration I did learn in my youth', he said laughing. 'Once, when the twittering of birds hindered my work and I closed the window, my mother became so angry with my foolish sensitivity that she slapped me around the ears and threw open the window again. That slap has been more useful to me than many an educational word.' Even working in the midst of the many boys had taught him to close off the outside world when necessary. The downside of his great concentration was his real professorial absent-mindedness. Fortunately, however, there were always loving persons to help him in this weakness. 'Professor, thou hast not yet put on thy hat', the clerks warned many times, when he left the laboratory, or 'Thy trouser leg is wrapped upward'. Or they phoned to remind him there was a thesis defense ceremony where they awaited him. The old maid, who had a great memory, was invaluable in remembering when exams and other important facts of the day were to take place. It happened that he once traveled to America and noticed in Rotterdam that he had forgotten his wallet with money. He could just return to Groningen, fetch it and reach the boat on time. Another

time in America in the train he had lost his wallet; after a desperate search the black conductor was consulted, 'Just look under your pillow, Sir,' was his advice. And indeed, that is where the wallet was!

The laboratory's clerks, whose numbers grew all the time, were completely devoted to him. They were filled with a reverent love and an unshakable faith. And that was no wonder, for he sympathized with them in their joys and personal difficulties, supported and counseled where life became a burden, had infinite trust, and was, as they all felt, like a father to them. In his laboratory the tidy, serene atmosphere, which Kapteyn brought with him, where ever he was, and which brightened the work, which was often endlessly mechanical and tedious, for them all. His radiant optimism helped them through many difficulties. Come, let us get started and then we will find out how to get on. We will survive.' And then, when the problem was faced with courage, the difficulties were solved automatically.

The afternoons were for contact with the outside world: his lectures, meetings, exams, his walks, including the regular Monday afternoon walk with two faithful friends Prof. Heymans, the philosopher, and the historian Prof. Boissevain. The threesome was a familiar appearance on the road to Haren every Monday. Heymans in the middle, the tall impressive figure in the peregrine coat with a distant view and his spirit well above the ordinary human fuss, Boissevain, small and thin, agile and full of lively interest, Kapteyn on the other side, slim and animated, enjoying the outdoors and the interesting things, that each of them discussed about their work and experiences, at the same time noticing and recognizing every bird call. They walked to Haren, a village an hour away from Groningen, and never changed that habit. For 20 years they persisted, and it was remarkable, they noticed, how seldom the infamous Dutch winter weather forced them to forego the walk. It did happen, especially in later years, that at the beginning of the walk they entered the pub 'De Passage' [the passing, transit], and sat there talking and then come home just as content. It was a beautiful friendship, full of mutual admiration and appreciation.

Ursul Philip Boissevain (1855–1930) was a professor of ancient history, Gerardus Heymans (1857–1930) professor of philosophy and psychology. The latter (see Fig. 8.3) was specialized in what he called 'special psychology', for which he conducted surveys among large numbers of test persons on the basis of questionnaires. He had set up a psychological laboratory for this purpose, where he had derived character traits of persons from a hundred or so biographies. It is said that a considerable amount of the mathematical work needed to process these lists and to find correlations between the character traits was done by the 'calculators' in Kapteyn's laboratory. This may very well be possible.

Heymans was also interested in differences between men and women. He researched this with questionnaires that he asked professors and lecturers at

Fig. 8.3 Gerardus Heymans, professor of philosophy and psychology. University Museum Groningen

Dutch universities to fill out; in these he asked about character traits such as individuality, ability for abstraction, memory, etcetera, and whether they felt predominantly male or female. He found no special result that could not easily be explained by prejudice. Maybe Kapteyn's staff contributed to this as well. Although Kapteyn and Heymans were good friends, they often had different opinions, but when it came to women's suffrage, they were on one line. Heymans lobbied more actively for women's suffrage, but Kapteyn was certainly involved in Heymans' circle of acquaintances who worked for women's emancipation. It is also remarkable that Kapteyn was the first astronomer in the Netherlands with a female Ph.D. student, Etine Imke Smid (1883–1960). She obtained her Ph.D. degree in 1914, unfortunately with a not very impressive dissertation. Henriette Hertzsprung–Kapteyn (Fig. 8.4) [1] continues her description:

Fig. 8.4 Henriette Hertzsprung–Kapteyn with her father. This picture has been taken from her biography [1]

Kapteyn suffered a lot from nervous pains in his eyes, for which he could not find any cure, but the defect healed spontaneously. He was forced to give up reading at night. This was a great sacrifice, but good friends helped him to cope with it and came to read to him regularly in the evening. Heymans did this for many years and both enjoyed these get-togethers. Initially they read articles from Kapteyn's research area, later on also other things, which interested both of them, mainly articles or brochures about the current problems in the scientific world. They discussed the problems, and since they were very different in disposition and temperament, there was a lot of ground for discussion, which always brought up new points of view and new questions. Both of them had the greatest respect for science and a firm beliefs in her sovereignty. [...]

Heymans and Kapteyn also studied other works. They took up Bacon's famous *Novum Organum* and even Boswell's *Life of Johnson*, that voluminous book that everyone knew by name, but which hardly anyone had read. Furthermore, they read many writings about Einstein's theory of relativity, which interested him very much, but for which he felt he could not really 'look through' it sufficiently. Prof. Ehrenfest, the physicist in Leiden, offered to come and deliver some lectures on it, for which he came to Groningen. Also philosophical articles were discussed, which, however, did not satisfy Kapteyn because of his strong sense of demonstrable reality, so that he could hardly 'get into' it. An exception to this formed exact psychological investigations, for which he always kept a lively interest. He followed Heymans' personal research in this field with great interest, and Heymans often made use of his always available help with mathematical formulations. This great helpfulness was one of his most characteristic qualities. His writings on the *Skew frequency curves* and on the *Correlation coefficients* were for example written for the benefit of Groningen biologists.

Sir Francis Bacon (1561–1626) was a British philosopher, scientist and politician and his *Novum Organum* is seen as the basis of modern scientific research based on experimentation. James Boswell (1740–1795) was a Scottish lawyer and writer, particularly notable for his biography of the British author, poet, playwright, etc. Samuel Johnson (1709–1784). Paul Ehrenfest (1880–1933), born in Austria, was professor of theoretical physics in Leiden. Gerardus Johannes Petrus Josephus Bolland (1854–1922), mentioned in the next paragraph, was professor of philosophy in Leiden.

Although Kapteyn was not impressed by pure philosophy, because by nature he was not sufficiently reflective, still made every effort to keep himself informed. For a year he attended a lecture series by Prof. Bolland, the renowned philosopher in Leiden, not because he was attracted to the personality and doctrine of this man; on the contrary, these repelled him, unbalanced and uncontrolled as they appeared to him, but because he wanted to get to know and try to understand, what captivated so many people and enraptured them. Temporarily Bolland's fiery eloquence captivated him, but all too soon he felt it changed into unmotivated outbursts and illogical twists, and his overconfident contempt for dissenters aroused Kapteyn's indignation. Levelheaded as he was, this inflammatory and uncontrollable temperament, this absolute subjectivity was as strange to him as it was unsympathetic. Science, according to him, who from an early age was accustomed to thinking in the way of natural sciences, had to be first and foremost objective; without objectivity, he believed, it was like a ship without a rudder, from which nothing good or permanent was to be expected, and could only confuse and obscure. Lessing's philosophy and classical tranquility attracted him much more. His wisdom was closer to him, and he resolved after his retirement when he hoped to devote himself more to literature, to study this philosopher thoroughly. A few days before his death he ordered Lessing's complete works. He has only seen the beautiful volumes and looked forward to reading them, but they remained unread. So much remained undone, which his retirement had promised him. Also Hudig, a young neighbor, who quickly became a friend, read to him regularly from geological and historical works, and those evenings were a great pleasure for both of them. Mrs. Kapteyn read him novels of their choice every day, and colorful magazine stories that amused him. They also worked together a lot, because gradually it became a habit for her to write the letters he dictated. And since his correspondence was very extensive, this was an invaluable help. And so the difficult years went by, until his eyes healed and he himself was master of his fate again.

Joost Hudig (1880–1967) was a chemical engineer, specialized in soil science. He worked at the 'Rijks-landbouwproefstation', the national facility for agricultural experiments in Groningen. He assisted Kapteyn in his activities for the Natural Sciences Society, and, as mentioned above, he later became Henriette's second husband.

When he was about 60 years old, he attended a music course to learn to understand this form of art better. He was not very talented musically, never played any musical instrument and did not have a beautiful voice. However, he had his own way of musical expression, which was called 'trumpets' at home. As soon as it was heard, Mrs. Kapteyn was irresistibly driven to the piano and she accompanied him, which created a rather original effect. One could hear him coming home singing all the time, even in the middle of his work he could suddenly burst into singing, mostly pieces from well-known sonatas and symphonies, which were familiar to him. He always wished to hear the same old familiar pieces, the new compositions had little attraction to him. But art itself, the unattainable, was for him a mystery, it filled him with a quiet, almost shy reverence and for an artist he felt the deepest awe. He followed a music course, given by Peter van Anrooy, the then conductor of the Groningen orchestra, which interested him exceptionally. Every Wednesday evening he attended the musical performance of the orchestra in 'de Harmonie', he immersed himself in the beauties of music and felt himself enriched all over again. His acquaintance with van Anrooy, which soon became a warm friendship, brought him closer to art. He found many parallels between science and art. Were not both in their highest form selfless and working for the ideal they had set themselves, searching for truth and purest expression? Insensitive to worldly fame and prosperity to give the highest that dwells in her? This is how he saw art as the sister of science.

Peter Gijsbert van Anrooy (1879–1954) was appointed conductor of the local orchestra of Groningen in 1905. In 1910 he became conductor in Arnhem and again later (in 1917) of the 'Residence Orchestra' in Den Haag, also known as the The Hague Philharmonic. The 'Harmonie' (see Fig. 8.5) dates from 1856. It featured a large concert hall with reputedly famous acoustics. After

Fig. 8.5 Part of 'de Harmonie', the Groningen concert hall, seen from the east (the Oude Kijk in 't Jatstraat). The building complex stretches out beyond the right edge of the photograph and towards the back, where the actual concert auditorium was located. University Museum Groningen

a fire in 1941 the building was renovated around 1990; it now houses the Groningen University Faculties of Law and Arts.

One beauty remained hidden to him, however, that was poetry. Not that he pitifully looked down upon it, as so many intellectuals do, but he lacked the sensitivity for it. Judge for yourself on the basis of the one poem he wrote, bestowed by him to the world. It was the poem he wrote when initiated in the Utrecht student society upon entering university. Now it has to be admitted that this period really is not the right time for high-level poetic expression, but his poem was undoubtedly the most un-poetical among those of his fellow students that went through initiation[1]:

> 'The initiation poem is obligatory
> for students to be initiated, I observe.
> So it seems to be the wisest for me
> To accept this custom without reserve.'

---

[1] Het groenversch is een eerstvereischte/ voor groenen, zo merk ik/ Het is dus zeker wel het wijst/ dat ik mij naar die gewoonte schik.

Is anything less poetic possible? Still it is interesting, since it shows characteristically his logical philosophical mind! Only epic poetry could charm him. His favorite song was the *Walthari Song*, that Germanic song of delightful primitiveness, of fighting and blood, of primeval people with primeval instincts. It grabbed him and aroused his enthusiasm. Many found this incomprehensible in him, but precisely his simplicity, that touch of primeval nature, which was undeniably his own, found satisfaction and 'Anklang' [German for appeal] with this thundering song of the powerful man in his unbridled passions. In many forms the beauty of life revealed itself to him: he had an unlimited ability to learn and enjoy. The first flowers and birds were a great happiness for him every spring. From walks, he always brought the first short-stemmed daisies, which he put on the table partly shyly and partly overjoyed. Every new bird he heard brought him in ecstasy and the first swallows were noted on the calendar. 'In my next incarnation I would like to be a swallow,' was his often expressed wish. They seemed to him the personified, carefree joy of life. Some other time he wished to become a dandy in a new life. The admiration for a well-groomed appearance always stayed with him. 'There is a lot of self-confidence in a elegant appearance', he often said. He never reached this ideal; the best he managed was a clean collar every day.

He could laugh irresistibly, from the bottom of his heart, especially when he told an anecdote – whether it was the first or the twentieth time he told it, it didn't matter – he laughed with the tears in his eyes, and the saddest of listeners forgot for a moment his black thoughts and laughed along, overcome by this gift and true merriment of the heart. A wise Frenchman once said these profound words:[2] 'The wise man doesn't laugh, he smiles.' Kapteyn knew better. [...]

He also did not condemn the cinema, as was customary in intellectual circles. He gladly went there with an open mind, looking for relaxation without asking for deeper values. Why always this destructive criticism of hopelessly superficiality and inferiority in life? Why does one have to cling so desperately to a life of superior values? Much more harmful is the deadly criticism that causes everything to wither away that it touches. One can after all put the draconian and sensational to the side, like one does with theater or literature. And then enjoy the unlimited technical and artistic possibilities that modern art offer.

Once, during the fair in Groningen, a tent was put up on the Ossemarkt, just in front of our house. Every evening we heard the most interesting sounds: gunshots, cheers, insane clapping of hands and we could not resist, had to go and see and enjoy the excitement. Sitting in the front row, my father and I watched a foolish display of stupid detectives who kept falling over and over again, creepy bandits singing a long song of 'The birds of the night' at the most critical moment, as well as the mildest, most improbable situations. And we laughed and clapped at so much folly and sang the song of the birds of the night for a long time

---

[2]Le sage ne rit pas, il sourit.

afterward. Here, too, the divine rhythm of the great, multifaceted life can reveal itself to those who have an eye for it.

8.4 Weersma, Yntema, van Rhijn

It took a long time before Kapteyn delivered his first Ph.D. student, Willem de Sitter in 1901. After that it took a long time again for number two to present himself. It is true that Cornelis Easton received his doctorate in 1903, but as said, that was *honoris causa*. In total Kapteyn, not counting Easton, only had eight Ph.D. students. Adriaan van Maanen (1884–1946) obtained his Ph.D. in 1911 in Utrecht with Albertus Antonie Nijland (1868–1936) as supervisor, but van Maanen actually worked in Groningen for some time on his dissertation (on the proper motions of stars) and in fact saw Kapteyn as his real teacher. With Adriaan van Maanen added, the number comes to nine. For Kapteyn's contemporaries in mathematics and natural sciences in Groningen, the average per professor was roughly one Ph.D. every three years. Kapteyn, who has been a professor for 43 years, even if Adriaan van Maanen is counted, does not get more than one Ph.D. student per almost five years. But when in 1908 his second Ph.D. student Herman Albertus Weersma (see Fig. 7.9) defended his thesis, he was far behind even on that schedule.

Herman Weersma's dissertation was about a new determination of the Apex 'according to the method of Bravais'. Weersma went back to the original formulation of the problem, using the spatial positions, velocities and masses of the stars, to determine the center of gravity of a system. This was not very practical of course, because the distance and spatial velocity of only a few stars had been measured; the mass of stars and the variation among them were largely unknown. So Weersma developed an adapted method, in which he took into account the missing data as well as possible and as a first approximation ascribed equal masses to all the stars. That was indeed an improvement on earlier work and he obtained a more accurate result, working with the care that Kapteyn expected of him. The knowledge of the position of the Apex (and the velocity of the Sun relative to the collective of stars in its vicinity) was, as we have already seen, an essential part of Kapteyn's program using secular parallaxes as a step towards determining the construction of the heavens.

Weersma was appointed on a permanent position as Kapteyn's assistant in 1908; he succeeded de Sitter when he moved to Leiden in that year. When Weersma left Groningen in 1912, his place (in 1913) was taken by Frits Zernike (1888–1966), who was later to win the Nobel Prize in physics for his invention of the phase contrast microscope. Zernike did not work for Kapteyn for very long; he obtained a Ph.D. in chemistry in 1915 (at the University of Amsterdam) and was subsequently appointed lecturer in theoretical physics at Groningen. His position at the Astronomical Laboratory was taken by Kapteyn's later successor, Pieter Johannes van Rhijn.

Others who worked on a dissertation with Kapteyn did so while not being employed as his assistant, but either at their own expense or on temporary appointments. Lambertus Yntema was Kapteyn's third Ph.D. student. He was born in 1879 in Friesland and obtained his Ph.D. degree in 1909. After this he became a teacher and director of secondary schools in Breda, Leeuwarden and Amsterdam. His dissertation concerned the total amount of starlight reaching us from the sky. In presenting his *Plan of Selected Areas*, Kapteyn had identified this as an important constraint in testing models for starlight distribution in space. After all, each model makes a prediction of the total number of faint stars, even for the ones that cannot be detected individually, even with a telescope. One can add up the light of all stars including the faintest, and that gives a value for the brightness of the background sky. And that must be in accordance with the observed value.

Measuring the brightness of the night sky—at a dark site, of course, and without the Moon being above the horizon—is no easy task. Kapteyn had experimented with it when he was in South Africa by holding up a sheet of white paper and illuminating it from behind him so that it was brighter than the sky, then turning the lighting down until it no longer contrasted with the sky and became a starless spot otherwise indistinguishable from the background sky. Together with Kapteyn, and with the help of the physics laboratory's workshop, Yntema designed two special setups to make an attempt (see Figs. 8.6 and 8.7). The way the first device worked was similar to Kapteyn's earlier experiments. In Fig. 8.6 we see a long, thin pole, which was pointed towards the sky and along which a lamp could move. In the figure it is just above its center of the pole, indicated with the letter 'M'. At the top left end was a small screen, illuminated by the lamp ('Q'). Just below it was a circular plate with a hole in it on a bracket placed slightly next to the pole ('S'), allowing the observer to look at the screen. In the screen 'Q' was also a hole, through which the night sky could be seen and compared to the brightness of the surface of the screen. To prevent scattered light, the part with screen 'Q' and picture 'S' was placed in a box that is seen in the picture to the left side on a stand. This box ('T') had two holes, the larger one to let the light of the lamp through, the smaller one for the observer to look through.

The measurements were carried out by Yntema in Borger, a village 36 km south-east of Groningen. His wife made all the notes. They were sitting on a flat roof, surrounded by a blinding fence. They checked the stability of the

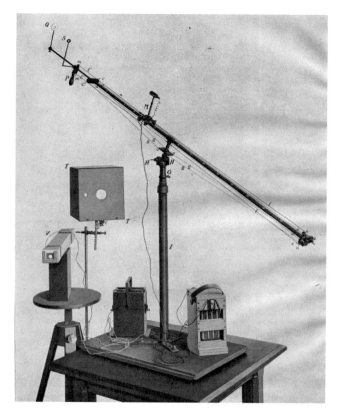

Fig. 8.6 Measuring device of Yntema to visually measure the brightness of the night sky. Kapteyn Astronomical Institute, University of Groningen

measuring construction with the device standing on the left on a stool ('V'). It contained an electric lamp and a frosted glass window. When put on a nearby roof it looked like a star and that was adopted as the standard. Using a Zöllner-photometer (see Fig. 2.6) the brightness of that artificial star was calibrated with respect to real stars.

The second setup Yntema used for his measurements was photographic (Fig. 8.7). Each box in the setup has a hole at the top to let the light of the night sky through and another hole at the bottom, where a photographic plate was mounted. Yntema usually used more boxes at a time than the three in the figure. He calibrated the device by pointing one of the boxes at the Moon.

The result of the measurements was somewhat disappointing. Yntema found that the difference in brightness between the Milky Way and its poles was 'surprisingly small'. Also, the brightness increased as one approached the horizon, whereas one would expect that, because of the longer path length through the Earth's atmosphere, the starlight would actually become less bright. Yntema

Fig. 8.7 Yntema's measurement setup to determine the brightness of the night sky photographically. Kapteyn Astronomical Institute, University of Groningen

concluded that there was undoubtedly a considerable amount of light coming from the atmosphere and the Earth itself; he called this 'Earth-light', and described it as a kind of 'permanent aurora'. But Yntema did not allow himself to be distracted by this. He tried to separate the contributions of starlight from the 'Earth-light' by estimating the relative brightness of various parts of the Milky Way itself and of its poles from the star counts determined by Kapteyn.

In fact the results were better than Yntema and Kapteyn could ever have suspected. In modern units we express such 'surface brightnesses' in magnitudes per square second of arc, as I already discussed in another context. To direct the thoughts: a night sky in a dark site on Earth, like a modern observatory, has a brightness, in the wavelengths that our eyes can see, of the order of 21 or 22 magnitudes per square arcsecond. This means that as much light comes from a piece of sky of 1 by 1 arcsecond as from a star of magnitude 21 or 22. Remember that the faintest stars visible to the naked eye are of magnitude 6, so such a star is one to two-and-a-half million times fainter. Yntema found (translated to these more modern units) that the terrestrial light in the zenith (the point right above the observer) was between 20.3 and 20.6 magnitudes per square arcsecond. He estimated the starlight in the same units as 20.4 in the brightest parts of the Milky Way to 22.6 in the Galactic poles.

Nowadays, Yntema's Earth-light is referred to as 'airglow'. It originates in the higher layers of the atmosphere as a result of processes in which molecules are ionized during the day by sunlight, i.e. electrons are being released, while they 'recombine' again during the night; and by interaction with cosmic rays, high-energy particles from the Sun or from interstellar space (produced in supernovae and the like). It is about one and a half magnitudes fainter than Yntema estimated. There is another component, which Yntema did not identify, and that is called zodiacal light. The name comes from the term zodiac; it is mainly found in the ecliptic, the orbital plane of the Earth and the other planets. It is sunlight that is reflected by dust in our planetary system. Its surface brightness is comparable to that of starlight.

Airglow and zodiacal light still make a good determination of the background starlight from Earth problematic, but now this can be done from space. Of course then there is no airglow, but still there is zodiacal light. Now, the dust responsible for this is more common the closer you get to the center of the Solar System – the Sun. A satellite that goes beyond the orbit of Mars does not really suffer from it anymore. The spacecraft Pioneer 10 and 11, for example, which went to investigate the planet Jupiter closely, traveled beyond Mars. They were launched in 1972 and 1973 respectively and passed Jupiter at the end of 1973 and the end of 1974. Pioneer 11 also flew past Saturn and NASA was able to follow these satellites for a long time after they had left the planetary system. On board these satellites were instruments that charted the sky from the slowly rolling spacecraft, after they had passed the asteroid belt. There is then much less dust and hardly any zodiacal light. So they could map the background starlight, ironically not in order to study it, but to be able to correct observations of zodiacal light from Earth. I did publish a study of this material in 1986, in which I used it to investigate the distribution of starlight in our Galaxy, just as Yntema and Kapteyn had intended long ago.

Yntema remarked at the end of his thesis that the result of his measurements was uncertain, but that a considerable improvement was possible if the experiment was repeated from a high mountain peak in a dark environment. This possibility arose during Pieter van Rhijn's doctoral research. As discussed in more detail in the next chapter, between 1908 and 1914 Kapteyn traveled annually to Mount Wilson Observatory in Southern California, and van Rhijn as a result of this spent some time there in 1913. There he repeated Yntema's measurements and the analysis thereof he made public in his dissertation in 1915, although a formal publication of the result in a scientific paper was not forthcoming until 1919 and 1921, probably because he had too much other work or wanted to perfect the analysis. Van Rhijn realized that part of the background light seemed to be related to the ecliptic and that this suggested that the zodiacal light, that was seen near the Sun just before sunrise or shortly after sunset, in fact extended over the whole sky. He corrected for it and thus came to a good agreement with the work of Yntema. In absolute terms: Yntema had found that the total starlight over the whole sky was that of 1350 stars of the first magnitude; van Rhijn arrived at 1440. Given the difficulty of the measurements, these results corresponded remarkably well. Compared to modern measurements, these values are about a third too high. Not a bad result at all.

Fig. 9.0 Kapteyn around 1910. Kapteyn Astronomical Institute, University of Groningen

9

Mount Wilson

If I have seen further…
it is because I have a bigger telescope than you.
Anonymous
Astronomers are only satisfied
when there's a telescope on every molehill.
Jan Bezemer (1938–2012)

We have seen above that George Hale, after building the largest refractor telescope at the Yerkes observatory, wanted to set up a large observatory in California on Mount Wilson. Mount Wilson is part of the ridge around the Los Angeles Valley, north of Pasadena, and overlooks the valley. Initially, Hale had an observatory in mind with telescopes for researching the Sun, but later he wanted to set up large telescopes there for nocturnal observations. One needed to study stars as well before the Sun could be understood. As early as 1894 Hale had persuaded his father, William Ellery Hale (1836–1898), a wealthy

---

Persiflage of Isaac Newton's quote *If I have seen further it is by standing on the shoulders of giants*. Compare also to the also anonymous *If I have not seen as far as others, it is because giants are standing on my shoulders*. Bezemer was a geologist and diplomat at the Ministry of Education, Culture and Science, who for many years represented the Netherlands in the Council of the European Southern Observatory ESO. He made this remark at a Council meeting in 2005 (it is first hand; I was president of that body at the time), when the Very Large Telescope VLT with four 8.2-m telescopes had just been put into operation in Chile, construction of the Atacama Large (sub-)Millimeter Array ALMA with 50 (later 60) dishes at an altitude of 5000 m in the Andes was well underway and plans were made to build a 50- up to 100-m optical telescope, the Extremely Large Telescope (ELT, now under construction with a 39-m aperture and planned for operation in 2025). Although he strongly and effectively supported all these efforts, he could not resist making this remark.

P. C. van der Kruit, *Pioneer of Galactic Astronomy: A Biography of Jacobus C. Kapteyn*, Springer Biographies, https://doi.org/10.1007/978-3-030-55423-1_9

Fig. 9.1 Mount Wilson Observatory. This illustration comes from Henriette Hertzsprung–Kapteyn's biography [1]

American who had made a fortune building elevators, to buy him a 60 in. glass disk to grind into a mirror for a telescope. The intention at the time was of course to put that telescope at Yerkes, but then there was no money to take the project any further. But George Hale succeeded to convince steel magnate Andrew Carnegie to first finance the Mount Wilson Observatory with its solar telescopes and next to build a 60 in. telescope there. At one-and-a-half meters, this was by far the largest telescope in the world, to be erected in the excellent climate of Southern California (Fig. 9.1).

Along with George Hale a large group of astronomers came from Yerkes to California, particularly Walter Sydney Adams (1876–1956). New collaborators were also attracted, including Frederick Hanley Seares (1873–1964), who would work extensively with Kapteyn. The construction of the 60 in. telescope, in which telescope designer George Willis Ritchey (1864–1945), who also came from Yerkes, played a crucial role, was already an enormous feat (Figs. 9.2 and 9.3). But in 1906 the local businessman John Daggett Hooker (1838–1911) spontaneously offered to finance a 100 in. telescope (2.5 ms!) for Mount Wilson. So even before the 60 in. telescope was completed, work was already underway to build a second, even larger giant telescope. Hale suffered more and more from depressions and nervous collapses, probably from pressure and the responsibility for these enormous projects.

Fig. 9.2 Lower part of the toll road up to Mount Wilson around 1904. Courtesy the Observatories of the Carnegie Institution for Science Collection, Huntington Library, San Marino, California [78]

Fig. 9.3 The tube of the 60 in. telescope being transported up the Mount Wilson trail. Kapteyn would have had to travel up this road also. Courtesy the Observatories of the Carnegie Institution for Science Collection, Huntington Library, San Marino, California [79]

Fig. 9.4 The 60 in. telescope at Mount Wilson Observatory. Courtesy the Observatories of the Carnegie Institution for Science Collection, Huntington Library, San Marino, California [80]

While the 60 in. telescope (Fig. 9.4) was being built, George Hale was looking for research programs for this giant telescope once it would be operational. He was a solar astronomer himself (today usually called a solar physicist) and was very impressed by Kapteyn, whom he had met in St. Louis in 1904. Kapteyn must also have talked to him about his *Plan of Selected Areas*. The Plan, as published in 1906, does not contain a commitment from Hale to participate. Nevertheless, Hale soon appears to have decided that Kapteyn's Plan was an ideal program for his 60 in. telescope.

9.1   Hale and Kapteyn

On January 9, 1907, George Hale (Fig. 9.5) wrote to Kapteyn. 'I am afraid that you have come to the conclusion that I am not interested in your *Plan of Selected Areas*. I do, however, have a great faith in that plan and consider it of the utmost importance.' And he asked Kapteyn to what extent the 60 in. could be useful for it. The construction of the telescope had been considerably

Fig. 9.5 George Ellery Hale, director of the Yerkes and of the Mount Wilson Observatory. This picture is from Hertzsprung–Kapteyn's biography [1]

delayed by the heavy earthquake of 1906 in San Francisco (where the telescope structure was produced). He suggested that Kapteyn come and visit once the telescope was ready and to make observations with the 60 in., and later with the 100 in. telescope. And in a letter from March of that year Hale promised that they can certainly work together.

The Hale Archives, kept at the California Institute of Technology (CalTech), contain a letter from Kapteyn dated 12 June 1907. It refers to a proposal Hale apparently had made to him the day before, which came as a complete surprise to Kapteyn. There is also a telegram from Kapteyn, dated 9/6/1907, which was stored in the Archives as if this meant September 6th. In fact, it is dated June 9th. In it Kapteyn confirms that he will be at the Hotel de l'Europe on Tuesday at 1 o'clock in the afternoon. Indeed, June 11th, the day before Kapteyn's letter, was a Tuesday. George Hale was traveling in Europe (in May there had been a meeting in Paris of the International Solar Union, founded in St. Louis) and he also visited Amsterdam; he had asked Kapteyn to meet him at his hotel. Hotel de l'Europe was not a cheap hotel then either, so Hale did travel 'like a gentleman'.

Hale's proposal was amazing. He suggested Kapteyn would accept a part-time appointment with the Carnegie Institution, which operated, among other things, the Observatory on Mount Wilson, and then come to California every year for several months to lead the implementation of the 60 in. telescope observational program for his *Selected Areas*. He was allowed to make use of the appointment as he wished; he was also allowed to do the work for it elsewhere, even outside America if there was a good reason for doing so at any time. One can imagine how Kapteyn must have felt during the train journey back from Amsterdam to Groningen. He now had the opportunity to finish his life's work; his Plan had become a spearhead of the new telescope, the largest on Earth and he was given the financial means to play a leading role in its implementation.

Apparently, according to Kapteyn's letter, Hale had offered him a salary of 1,500 dollar a year. That sounds like quite a sum, but in his reply Kapteyn suggests raising it to 1800 dollars! I quote [Kapteyn's wording is not very transparent, but the intention is clear]:

> Not that I will not accept if there are fundamental objections and then accept the earlier amount. If I were rich enough I would undoubtedly prefer not to receive any salary at all. But now that I am not in the circumstance to cherish such a luxury, it would be a fine thing for me to be free of financial worries; which means: free from the need to give extra lectures, to take external exams and so on in order to be able to provide for the claim on a professor's salary by my three out-of-home children. In short, the requested increase should allow me to dedicate all my free time to astronomical research.

The 1800 dollars in 1907 represents a purchasing power of about 50,000 € now. That would now be almost half a year's salary of a senior professor. But it is likely that Kapteyn would have to pay his travel and other extra costs out of the Carnegie allowance as well. George Hale agreed on June 15 (now from London) and even proposes to raise the salary even further to 2,000 dollars if Kapteyn's trip to America takes up too much of his resources.

9.2 First Visit, 1908

Kapteyn's first visit to Mount Wilson took place in 1908; the intention was that it would coincide with the commissioning of the 60 in. telescope. Kapteyn left the Netherlands on September 12th. However, the final construction of the telescope turned out to be more difficult than expected, and he spent most of his time at Mount Wilson waiting for the construction to be completed. He lived at the observatory itself. Mount Wilson is a mountain just north and a bit east of Pasadena, the town where the offices of the observatory are located. The way up, the so-called Mount Wilson Toll Road, started in Altadena, 10 km from Pasadena at an altitude of about 300 m above sea level. The road climbs to the top of Mount Wilson, which is at 1740 m, over a distance of 16 km. The trip was usually made with a wagon pulled by mules or horses, or sometimes on horseback. At the time of Kapteyn's first visit, the road had just been widened to about 3 m, but it must have been quite an adventurous and demanding undertaking (Fig. 9.3).

The program for which the 60 in. telescope would be used consisted of two parts. The most important and urgent part for Kapteyn was the systematic photographing of the fields in the *Selected Areas* in order to obtain counts of stars to fainter brightnesses than previously possible. To sketch the perspective the following comparison. The catalogs of stars that existed until then, such as the *Bonner Durchmusterung* in the north and the *Cape Photographic Durchmusterung* in the south, covered the entire sky; the faintest stars in these catalogs had a brightness of about magnitude 9.5, about 25 times fainter than the naked eye can see. Across the whole sky that was about 800,000 stars. The *Carte du Ciel* intended to catalog all stars to magnitude 11, so one hundred times too faint for the human eye. But to study the distribution of stars in space, it was necessary to go much further. The collaboration with Harvard and Edward Pickering, which would produce photographic recordings of the *Selected Areas* across the entire sky from two hemispheres, would lead to a catalog with all the stars up to about magnitude 14. But for the study he had in mind, Kapteyn realized only too well that even this was not enough. That is why the works with the 60 in. at Mount Wilson was so fundamental, because the aim—at least for the part of the sky that can be seen from California—was to complete the counts in the *Selected Areas* to magnitude 18 or 19, which is a hundred thousand times fainter than the eye can see. For comparison: the Sun would have an apparent magnitude of 18 at a distance of 4.3 kpc. And so the Mount Wilson program was crucial to Kapteyn.

The second part of the program concerned the determination of radial velocities of stars. This requires a large telescope, because the light is dispersed in wavelength and even with the 60 in. this only worked for relatively bright stars. The Plan itself did not involve radial velocities, because the *Selected Areas* contained much too few stars that were bright enough to obtain absorption-line spectra and measure the shift of the spectral lines. So this part of Kapteyn's program involved bright stars all over the sky. Walter Adams in particular focused on this kind of work. Frederick Seares was the one who was most closely involved in photographing the *Selected Areas* and who was in charge of the daily management of this work at the telescope. A deep friendship developed between Kapteyn and Seares.

So the first trip from Kapteyn to California was only a moderate success. Eventually he had to extend his stay because of the delay in the construction of the 60 in. The first photographic plates for Kapteyn's program were only taken in December (probably on the 12th).

He was able to visit his son both on his way to California and on the way back. He had settled in Denver, Colorado, after finishing his mining studies. Kapteyn also visited the Newcombs in Washington, where he also met Robert Simpson Woodward (1849–1924), the director of the Carnegie Institution, who financed his position as a Research Associate. Kapteyn left New York just before the New Year.

9.3   Kapteyn Cottage

Kapteyn did not like to travel without his wife. In Hertzsprung–Kapteyn's biography [1] she tells how he had great difficulty making contact with fellow passengers on board the ship and that he felt uncomfortable in their company. On later visits to America, Kapteyn did travel with his wife; she came along for the first time in 1909 (see Fig. 9.6). He immediately felt much more at ease. Mrs. Kapteyn's correspondence with Mrs. Newcomb in Washington shows that they had first visited Edward Pickering in Harvard. It was known that Simon Newcomb was seriously ill and of course the Kapteyns were very concerned. Mrs. Kapteyn apologized for not being able to come straight away, but promised to visit Washington on the way back.

Just like Kapteyn's trip to St. Louis, this one and the next trips to California were no sinecure. The crossing directly from Rotterdam to New York took about ten or eleven days. However, the Kapteyns often traveled via England and then visited the Gills in London. Usually they did not go directly to California upon arrival in the United States either, but visited astronomers in Harvard and Yerkes or other places on the eastern parts of the USA before traveling to the other side of the continent; they also visited their son in Colorado on the

Fig. 9.6 Kapteyn and his wife aboard a steamship bound for the United States and Mount Wilson. This picture is from Hertzsprung–Kapteyn's biography [1]

way. A direct train journey from New York to Los Angeles already took about four or five days (in 1876 there was a record of 83 h and 39 min, but of course that was not a regular scheduled journey and around 1910 this was probably still faster than what was being offered in the timetable); with all the detours they took, we may safely assume that on the way to Pasadena as well as on the way back they were traveling maybe as long as two to four weeks.

Upon their arrival in Pasadena the Kapteyns heard that Simon Newcomb had died. The trip of 1909 is reasonably documented in letters between the ladies Kapteyn and Newcomb. On July 4, Elise Kapteyn wrote that they had arrived in America. From Mount Wilson she wrote that they planned on their way back to travel first to Williamsbay (i.e. visiting Frost at Yerkes), arriving there on the 27th or 28th of September and to visit the widow Newcomb on the 29th or 30th. Kapteyn wrote to George Hale on October 9th that they were about to leave the United States and later that they had returned home

Fig. 9.7 Kapteyn and his wife at the tent on Mount Wilson, where they lived during their visit in 1909. Kapteyn Astronomical Institute, University of Groningen

on October 25th. That means that they had been away from home together for about four months.

The Kapteyns lived on Mount Wilson, close to the telescopes. In 1909 this was still in a tent (see Fig. 9.7). Their supplies had to be brought from below; if there was a need for something, their request was taken down and the items were brought up as soon as possible. It is unlikely that they went to Pasadena on their own, because the journey along Mount Wilson Toll Road took quite some time. The tent would not have been very comfortable. A lot of the daily activities had to be done in the open air, which generally was not a big problem because of the climate. They did have problems with flies and the like. But after that first trip, a special small house was built for them, which was ready for them when they arrived in 1910 (see Fig. 9.8).

It was a big improvement. In Hertzsprung–Kapteyn's biography [1] the cottage is described as a big surprise, which the Kapteyns found waited for them upon arrival. In the correspondence between Hale and Kapteyn, however, Hale had already at some stage reported that its construction was progressing satisfactorily. It remained a small accommodation with a minimum of facilities. It had no running water or a bath. There was no desk or workplace for Kapteyn and it is likely that he had to rely on the facilities of the 'Monastery', the home of the Pasadena astronomers who came to use the telescopes and where presumably Kapteyn had stayed during his visit in 1908. There was heating, which was

Fig. 9.8 The Kapteyn Cottage on Mount Wilson. This picture is from Henriette Hertzsprung–Kapteyn's biography [1]

a necessity during the winter months, but with a wood-burning stove. There was a small kitchen. The porch looked westward (see Fig. 9.9) and offered a beautiful view over the valley and Los Angeles; on clear days the Pacific Ocean was visible in the distance. The Kapteyn Cottage still exists, although refurbished and expanded. The next section of Hertzsprung–Kapteyn's biography [1] gives an impression of the stay of the Kapteyns on Mount Wilson.

> The Kapteyn Cottage, as it is named to this day, their American home for many years, quickly became as dear to them as their Dutch home. It was built on a lovely spot, shaded by gnarled oaks and ancient pines, and great white yuccas stood like giant bouquets around the house. The view was great, over the mountains and the canyon, the deep, dormant ravine of Western America, over the valley with its many cities, which in the evening lighted up like star clusters, increased the feeling of tranquility high above, but at the same time gave the lonely the feeling that the living humans were near. And in the distance the majestic water surface of the Pacific. The peaceful silence that prevailed there, far from the hectic world, was still doubly sweet to them after the tiring and eventful journey. Deer came, peeking curiously, squirrels rushed from tree stem to stem, and the chickadees, the tits of the West, twittered there all day. Lovely were the cool nights, and Mrs. Kapteyn set herself up for the night on the porch, the large veranda, which covered the whole width of the house. One evening she saw something dark on her bed. It turned out to be a nest of young squirrels, for whom the tender tired mother had chosen a soft spot to come into the world. Idyllic, but also a bit creepy, she thought. And she looked for another place to sleep that night.

Fig. 9.9 The present (2013) view from the veranda of the Kapteyn Cottage on Mount Wilson. Photograph by the author

Kapteyn missed the heather, which he loved so much, and he decided to bring it in. The next year he traveled with a box of heather as the most precious piece of luggage, which had his greatest attention. Once it was left behind in a train compartment, because of unexpected luggage shifts, but he telegraphed to everyone he could think of and it was delivered back to him. It arrived undamaged on the mountain and was planted near the house, but the heather could not catch on in spite of all love and care. And this wish he had to give up.

We have to realize that although Kapteyn received permission from the University of Groningen to make these trips, he had to make up for the teaching he would have done in the remaining months of the year. Nothing came for free. In addition, Kapteyn felt uncomfortable as a paid employee of the Carnegie Institution. According to Henriette Hertzsprung–Kapteyn's biography [1], he wrote to his friend Boissevain in 1913:

I don't think I will be able to continue with these California trips for many years to come. It would be the nicest, prettiest position imaginable – if I were not paid for it. Now that I am, I sometimes have the feeling: do I deliver value for money? I would still answer yes, if I could simply set the people here to work to do what I think is necessary. But of course I only have an advisory role – advice that is listened to, but still – the whole establishment is really a 'Sonnenwarte', an establishment for solar physics, and I represent the other part. The employees all belong to the other side. And yet again, they have done very, very much according to my wishes. That is the 'maladie du doute' [disease of doubt], deep in my character.

It is clear from correspondence that George Hale had a completely different opinion about whether Kapteyn did enough. I'll come back to this later when I describe how Kapteyn wanted to end the contract when he no longer could visit Mount Wilson. The position of Hale and the Carnegie Institution was that Kapteyn worked for them and that it did not matter where he did that.

9.4   Dust

We have seen above that Kapteyn, in his attempts to determine the distribution of stars in space, had to make three assumptions. That of the lack of a preferred direction in the velocity distribution of the stars had already been proved wrong by his discovery of the Star Streams. But he was able to take this into account in the analysis. He never had to give up the assumption that the intrinsic luminosity (the absolute magnitude) of stars was distributed the same everywhere in space; especially for the stars that Kapteyn used, which are generally not that far from the Sun, that assumption is not bad at all. But what worried him for years, and which also explains why his model turned out to be partly incorrect, was the assumption of the absence of scattering of light by particles of dust in space (usually summarized as 'absorption', although it is that only partially). It is especially this particular issue on which he spent much time in the early years of his visits to Mount Wilson.

It was not the first time he had made a study on the subject. Earlier, he had addressed the issue also in response to an article by Edward Pickering in 1903 about the distribution of the stars. Pickering thought that if you looked at fainter stars, the number of stars across the whole sky grew at a slower rate than what was expected 'theoretically'. This theoretical value is then the increase in the simplest case that all stars are intrinsically equally bright and uniformly distributed throughout space, i.e. their density—the number of stars per unit volume such as cubic parsec—does not change with distance from the Sun. The theoretical derivation is explained with some inevitable algebra in Appendix A.6. Pickering stated that this deviation from the theoretical value was probably the result of absorption (or scattering) of starlight in space. But this had to be considerable in order to explain the numbers observed. A similar suggestion was made in 1904 by George Comstock of the Washburn Observatory. He had used Kapteyn's method of estimating secular parallax distances for various groups of stars, and had come to the conclusion that stars that are faint in the sky are only partly that because they are further away, but for a large part because they have intrinsically less luminosity. He also suggested that absorption of starlight could be the cause of this. Kapteyn had commented on this during his presentation in St. Louis, stating that this conclusion was far from unique.

Kapteyn picked up his earlier work on the spatial distribution of the stars (see Sect. 6.6). He suspected that one could explain some of the anomalies found by the fact that you saw another part of the stellar distribution at greater distance, namely the ones that are intrinsically brighter. But to prove that this was indeed the cause, one had to have more information about the distribution of intrinsic luminosities, the so-called 'brightness curve'. The models that Kapteyn was able to construct for this were not very accurate, but indicated that it was likely that the results of Comstock and Pickering could be explained at least partly by this brightness curve and partly by the absorption of starlight, but that the latter was much less than these two astronomers had found. It was clear that this was a problem and that the absence of absorption, the assumption of a completely transparent space, was unlikely to be realistic. For Kapteyn's work, this presented a fundamental uncertainty and therefore a major problem. Kapteyn also disagreed with the conclusions of Pickering and Comstock, because models that assumed a high absorption of starlight resulted in a distribution of stars in space with a large peak near the Sun. That the Sun occupied such a special position was not acceptable to him, because the Sun was after all supposed to be an average star among the stars. Since Copernicus we are no longer the center of the Universe.

Before I discuss Kapteyn's work in this area during his first years on Mount Wilson, I will first discuss the current understanding. Figure 9.10, shows the constellation Orion.[1] In this picture we see not only stars, but also more extensive bright features. We see those particularly near the stars in the 'belt', which are the three stars in a row in the middle, and at the end of the downward pointing row below, which is called the 'sword'. This latter object is the Orion Nebula, which is shown in more detail in Fig. 9.11. There is also a large circular structure in the picture, at least part of which is called Barnard's Loop. The latter is probably an expanding shell that lies further from us than the Orion Nebula, and is left over from a supernova explosion at the end of a heavy star's life.

The Orion Nebula is an area of gas and dust. It is lit up by a group of young stars, some of which are very heavy and hot. They produce a lot of ultraviolet light, i.e. a lot of energy, and that radiation ionizes the hydrogen (and other elements too, but hydrogen is the most common element), i.e., it releases the one electron that is around the hydrogen nucleus, a proton.

---

[1]The name Orion is pronounced with the accent on the second syllable 'ri'.

Fig. 9.10 Photograph of the constellation Orion. The three stars on a row in the middle form the so-called 'belt'. The structure beneath it is the 'sword'; its end is nebulous; this is the Orion nebula of Fig. 9.11. From www.AlltheSky.com [81]

Electrons and protons attract each other. Protons (or atomic nuclei in general) and electrons can therefore combine again to form an atom, releasing energy in the form of light (this is called recombination). This then happens at very specific wavelengths and can be seen as definite emission lines in the spectrum. We see such light in the Orion Nebula. Because such nebulae are mostly ionized hydrogen they are called HII-regions, HII being (singly) ionized hydrogen (HI is neutral, unionized hydrogen).

However, not all structure we see in the Orion Nebula reflects the spatial distribution of gas. There is dust, also between us and the nebula. The separation between the large Orion Nebula and the small one on the upper left is an example of this.

Not all luminous nebulae are areas of hot gas like the Orion Nebula. In Fig. 9.12 we see the Pleiades, which we saw also in Fig. 7.11, in which the Hyades and the constellation Taurus were illustrated. The stars in this Pleiades cluster have quite a lot of dust around them and the small particles of dust

Fig. 9.11 The Orion Nebula is an area of glowing gas around a group of young, hot stars. It can be seen with the naked eye at the end of the 'sword' (see Fig. 9.10). European Southern Observatory [82]

reflect the light of the star. That is why the nebulous light in the Pleiades is also concentrated towards the stars. These nebulae are called reflection nebulae. One can easily tell the difference, because the spectrum of reflection nebulae has the same pattern of dark absorption lines as the star, and not the bright emission lines that the Orion Nebula has. We have to think of small dust particles with dimensions from much less than 1 micron (one micrometer or one thousandth of a millimeter) to 10 or more microns. They scatter light because their dimensions are comparable to the wavelength of the light. The wavelength of the optical light that we see with our eyes is about 0.5 micron. Interstellar dust consists mainly of magnesium- and iron-rich silicates, graphites (with a lot of carbon) and simply ice. At their surfaces, compounds (like carbon monoxide, but also alcohols and complex organic molecules) are formed by the action of starlight, which permeates interstellar space. Actually, the dust not only scatters the light, but also absorbs it, so that the energy of the light raises the temperature of the dust particle.

Because the light is scattered and absorbed by the dust, we see it as dark spots against bright backgrounds. A nice example of this can be found in the Horsehead Nebula, which is also found in the constellation Orion (see Fig. 9.13). The bright matter behind it is ionized gas by young, hot stars. The cool dust that lies further away from those stars becomes visible because it

Fig. 9.12 Clouds of dust in the Pleiades. This cluster can also be seen in Fig. 9.10, which shows the constellation of Taurus and the Hyades. The light of the stars is scattered by the cold dust, so that it becomes visible. Space Telescope Science Institute [83]

scatters the light from the gas behind it, so that it does not reach us. Another nice example of a dark cloud shielding starlight is the Coalsack visible to the naked eye near the constellation Southern Cross (see Fig. 3.5).

A very important aspect of the shielding of light by dust in interstellar space, also called 'interstellar extinction', is that its effect strongly depends on wavelength. Red light is scattered much less than blue light. This has to do with the size of the dust particles relative to the wavelength of the light. An excellent illustration of that effect is Fig. 9.14. We see a dark cloud, called Barnard 86, in front of a background of stars. In the picture above on the left, which was taken in blue light, we see the effect at its strongest. In visual light between blue and red in the top center the effect is slightly less; in red light, on the right, it is still there, especially in the middle. Only in photographs with longer wavelengths, in the near-infrared (which we cannot see with our eyes), do we notice that we can see better and better through the dust cloud. If we go from right to left in the bottom row of images, the wavelength increases from 1.25 μm (micron) to 1.65 μm to 2.16 μm. At the last wavelength we can still see that there are fewer stars to be seen through Barnard 68 than next to it, but we do see many more stars than in the optical.

Fig. 9.13 An image of a small area in Orion, just below the leftmost star in the belt (see Fig. 7.11). In this picture the north is to the left, so it is rotated by $90°$ anticlockwise. To the left of the middle the so-called Horsehead Nebula. This is a cloud of small dust particles that absorbs the light of the bright nebula behind it, which is hot gas like the Orion Nebula in Fig. 9.11. Space Telescope Science Institute [84]

9.5   Interstellar Extinction

None of this was known to Kapteyn and his contemporaries. Kapteyn wrote two very important papers on this subject, which he published in 1909. Because of his association with the Mount Wilson Observatory and the Carnegie Institution, he had begun to publish mostly in English; if it did not concern publications with many tables and observation data (which he reserved for the *Publications of the Astronomical Laboratory at Groningen*), he mainly submitted his papers to American journals such as the *Astronomical Journal* and especially the *Astrophysical Journal.*

In the first article Kapteyn looked at clues for interstellar material in spectra of stars. This concerned both spectral lines of atoms (or in principle also molecules), as well as possible effects of a general nature. For example, Antonia Maury, who we have come to know as one of the authors of the classification scheme of stars that culminated in the *Henry Draper Catalog*, had noted that spectra of stars sometimes appeared to show a general sign of absorption at blue wavelengths ('shorter than 4307', and that must be Ångström[2]). Kapteyn dis-

---

[2]Astronomers still often use this old unit for wavelengths, named after Swedish physicist Anders Jonas Ångström (1814–1874). It is one ten millionth of a millimeter or 10^{-10} m. So 10,000 Å is one μm (micron), and 10 Å a nanometer.

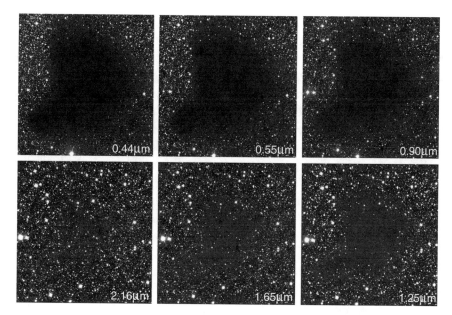

Fig. 9.14 Images of a so-called dust globule at different wavelengths. Above from left to right in the optical light at blue, visual and red wavelengths. Bottom at infrared wavelengths, increasing from right to left. This figure is produced by the European Southern Observatory in Chile. The pictures at the top have been produced with one of the four 8.2 m telescopes of the Very Large Telescope on Paranal, at the bottom with the New Technology Telescope on La Silla [85]

covered that stars that showed this phenomenon generally had smaller proper motions and thus statistically were at larger distances than stars without it. How much extinction there was, i.e. how many magnitudes the starlight had become fainter, could not be deduced.

The second article was much more specific. Here Kapteyn looked at the colors of stars. He postulated that scattering of starlight was probably strongest in blue, and that stars at larger distances from us would then be systematically redder. Kapteyn assumed that the mechanism involved was the so-called 'Rayleigh scattering', so named after Lord Rayleigh, John William Strutt, 3rd Baron Rayleigh (1842–1919). This mechanism explains why the light of the clear sky is blue. The scattering is very strongly wavelength-dependent, namely with the fourth power (so at two times shorter wavelengths it is $2^4 = 16$ times stronger). However, we now know that this kind of scattering is from atoms or molecules (in the example of the blue sky in high layers of the atmosphere). In interstellar space the scattering is from small dust particles. The wavelength dependence is therefore not the same as that of Rayleigh scattering, but Kapteyn did not know that.

What Kapteyn had at his disposal were measurements of the apparent brightness of stars at photographic and at visual wavelengths. Photographic refers to the wavelength sensitivity of the first photographic emulsions. It is relatively blue light with a central wavelength of around 4300 Å, but more than 1000 Å wide. Visually refers to the wavelength sensitivity of the human eye, whose central wavelength is close to 5400 Å. By comparing these two magnitudes, he could get an impression of the color of the light of a star. Kapteyn found that the color of stars depends on its spectral type. But if he included the effect of distances he also found an effect, so according to his analysis there were indeed effects of extinction. However, the change of color in itself does not yet tell us what the total amount of absorption is. For that, Kapteyn assumed that it was Rayleigh scattering and that it depended on the wavelength with the fourth power. He concluded that the 'loss of light' over his unit of distance of 32.6 light-years (10 pc) was 0.01 magnitude in the photographic, and 0.005 magnitude in the visual. Not a bad result, because the modern values are on average 0.013 and 0.010, respectively. He quickly published this result in the *Astrophysical Journal.*

However, Kapteyn discovered very quickly that he had made a fundamental mistake. Instead of calculating average distances, he had taken average parallaxes. Of course, a distance r is the inverse of the parallax p. So the distance is proportional to $1/p$. But it is a property of numbers that the average of a number of values of p is not the same as the inverse of the average of the values $1/p$. He therefore quickly sent an erratum to the editors of the journal. This changed his conclusion: the extinction turned out to be twice as small than he had previously concluded.

In 1914 Kapteyn returned to the question of extinction. In an article he published that year, he did not present a new analysis, but discussed the available literature. In the meantime, more was known about the nature of stars; in particular, it was now known that the intrinsically brightest stars were intrinsically bluer, and the intrinsically fainter were redder. So apparently fainter stars would on average be redder. This complicated the analysis. It boils down to the fact that the effects of a relationship between the color of a star and its absolute magnitude or luminosity, and reddening by dust in space were difficult to separate. Pieter van Rhijn studied this subject again in his dissertation of 1915. He showed that change in color of a star if it were due to extinction was 0.00195 ± 0.0003 magnitudes per 10 pc, about half of Kapteyn's value. Kapteyn and his colleagues still did not know whether this was really a matter of interstellar extinction.

This changed in 1916 with the publication of a study of the globular cluster Messier 13 (M13, see Fig. 9.15). Globular clusters belong to the halo of our

Fig. 9.15 The globular cluster Messier 13. From Wikipedia [86]

Galaxy. This is the more or less spherical distribution of the oldest stars, of which a small part still is in globular clusters. All stars in such a cluster have about the same age, which is between 10 and 12 billion years. Globular cluster M13 is seen on the sky outside the Milky Way (about 40° from it), and it is not part of the disk of the Milky Way Galaxy. Then the line of sight towards it is, as we now know, virtually free of extinction, because the dust is concentrated in a thin layer in the disk of the Galaxy. Harlow Shapley (1885–1972), who then worked at the Mount Wilson Observatory, determined the colors of no less than 1300 stars in the M13 cluster. Color means the difference in magnitudes of a star between a bluer and a redder wavelength. He estimated the distance to be a bit greater than it actually is according to modern determinations, namely 10 kpc rather than 6.8 kpc, but that does not fundamentally change his discussion. His conclusion was that if extinction was as large as Kapteyn had derived, the bluest stars should have colors a few magnitudes redder than the stars in the solar neighborhood. So space had to be transparent. His upper limit was an extinction of 0.0001 magnitude per 10 pc, at least ten times less than Pieter van Rhijn had found in that same year.

This observation convinced the whole astronomical world that interstellar extinction was unimportant and could be neglected. It was not until 1930 that Swiss-born American astronomer Robert Julius Trumpler (1886–1956) showed that the distances of star clusters in the Milky Way, as determined by apparent magnitude and using spectral types, resulted in values for the diameter (in parsec) of those clusters that systematically decreased with distance from the Sun. The diameter from colors is affected by extinction and that from spectral types is not. So there had to be extinction, and from that moment on everyone was convinced of that too. Of course, it then also became clear that the dust responsible for the extinction is concentrated in a very thin layer in the Milky Way. Kapteyn never included extinction anymore in his studies after 1916; Shapley's argument was sufficiently convincing. And not just for him; it was universally accepted.

9.6   The Solar Union Congress, 1910

During the stay of the Kapteyns on Mount Wilson in 1910, a large meeting was held of the International Union for the Cooperation in Solar Research. There was a lot of interest, partly because astronomers from all over the United States and Europe wanted to see the giant telescope and the enormous solar towers. Although it was a congress on solar research, Hale was actually planning to talk about coordinating solar research with that of stars, so the congress was an incentive to bring the two areas of research together. There were participants from both areas, with the result that most of the top astronomers from both America and Europe were present, as well as the directors of virtually all the influential observatories on both continents.

From the American institutes were in attendance Edward Pickering of Harvard, Edwin Frost of Yerkes Observatory, William Wallace Campbell (1862–1938) from Lick Observatory (not far from San Francisco), Frank Schlesinger (1871–1943) from the Allegheny Observatory (Pittsburgh), later director of Yale Observatory (New Haven, Connecticut), Edward Emerson Barnard (1857–1923) from Yerkes, Vesto Melvin Slipher (1875–1969) from Lowell Observatory near Flagstaff, Arizona, Henry Norris Russell—known from the Hertzsprung–Russell diagram—from Princeton (where he was to become director in 1912), Armin Otto Leuschner (1868–1953) from Berkeley. In fact, without exception, these were leading astronomers and astrophysicists, who investigated stars and nebulae and had little to do directly with solar research.

European participants were Frank Watson Dyson (1868–1939) then Astronomer Royal for Scotland, not long afterwards director of the Royal Greenwich Observatory, Jöns Oskar Backlund from Pulkovo near St. Petersburg, Herbert Hall Turner from Oxford, Karl Schwarzschild of the Sternwarte Göttingen, Johannes Robert Rydberg (1854–1919) of Lund Observatory in

Fig. 9.16 The participants of the congress of the International Union for the Cooperation of Solar Research on Mount Wilson in the summer of 1910. The building is shielded
from the Sun with a large tent. Courtesy the Observatories of the Carnegie Institution
for Science Collection, Huntington Library, San Marino, California [87]

Sweden, Karl Friedrich Küstner from the Bonner Sternwarte. They too were
astronomers who mainly studied stars. But also Marie Paul Auguste Charles
Fabry (1867–1945) from Marseille was there, together with Jean-Baptiste
Alfred Perot (1863–1925), together the inventors of an instrument to accurately measure wavelengths of spectral lines. And from Europe also the
renowned researchers of the Sun Henri-Alexandre Deslandres (1853–1948)
of l'Observatoire de Paris en Annibale Riccò (1844–1919) of Catania Osser-
vatorio in southern Italy.

Figure 9.16 shows all participants and Fig. 9.17 some key persons. A report
of the congress and an identification of all participants can be found in an
article by Herbert Couper Wilson (1858–1940) [88], who is known as editor
of two American astronomical journals. The meeting was held on top of Mount
Wilson in a wooden building on the Observatory, known as the 'museum'. To
keep the temperature acceptable, a large canvas tent cloth was stretched over
it (see Fig. 9.16).

The almost eighty participants had first (on Monday August 29) visited the
offices and the workshop of the Observatory in Pasadena. Many of them had
attended a meeting of the Astronomical and Astrophysical Society of America in
Harvard before this meeting (from 17 to 19 August) and had come to Pasadena
from Chicago in two special train wagons,[3] visiting Lowell Observatory in

---

[3]The report says that forty persons were transported with two 'cars'. I assume they were train wagons.

Fig. 9.17 The persons in the center in Fig. 9.16. From left to right in the front row: Walter Adams, Edward Pickering, George Hale, Mrs. Kapteyn and Williamina Fleming. Kapteyn is behind her. From the Observatories of the Carnegie Institution for Science Collection at the Huntington Library [87]

Flagstaff and the nearby Grand Canyon along the way. In Pasadena they saw the primary mirror for the 100 in. telescope under construction, which was being ground into the right shape. In the afternoon there had been a garden party, at the home of George Hale and his wife. The Kapteyns had been invited there as well, but Kapteyn had written Hale in Pasadena by letter from Mount Wilson that this was too time-consuming a trip and had apologized himself and his wife. He wrote that otherwise he would have to give up the idea of being able to finish 'to some extent' his work.

The trip up the mountain on Tuesday was done in two groups. Most of the guests took the Mount Wilson Toll Road with six-seater coaches. A smaller part took an older path, the start of which was reached with a trolley (a kind of electric tram). Then they traveled by horse and some even on foot. The whole journey took about eight hours, because both the walkers as well as the horses had to rest regularly. On top of Mount Wilson was a hotel, which had been opened for tourists in 1905, where the rich could escape the hot summer temperatures of Los Angeles and its surroundings on weekends. However, about twenty astronomers slept in tents because there was not room for everyone in the hotel.

The congress started on Wednesday and lasted until Friday. George Hale was only present on the first day because of his health. On Wednesday and Thursday evening there were special lectures; Kapteyn took care of Wednesday's lecture. He talked about the research he was doing at that time and concerned the distribution of what he called 'helium stars'. I will discuss Kapteyn's work later.

The Solar Union conference ended with the return from the top of Mount Wilson to Pasadena. It took considerably less time than the trip up (it took now only about four hours) and in the evening there was a big dinner, offered by George Hale and his wife, with over a hundred persons, conference participants and special guests. It turns out that most of the participants, since they were in California anyway, took the opportunity to visit the Lick Observatory, just south of San Francisco. The European participants at the conference on Mount Wilson must have been away from home from early August to early October.

9.7 Results of the Visits to Mount Wilson

The annual trips of the Kapteyns to California have been of great importance. In the first place for the work on the *Plan of Selected Areas*, and in particular the exposing of plates of all (eventually 139) northern *Selected Areas* that were observable from Mount Wilson. The first few years were necessary to prepare the whole thing; most of the photographic plates were taken between 1910 and 1912. The staff members Harold Delos Babcock (1882–1968) and Edward Arthur Fath (1880–1959) took the plates. The most important staff member, however, was the aforementioned Frederick Seares (Fig. 9.18). His share was much more than just the coordinating of the work when Kapteyn was not on Mount Wilson. In the end the work had to result in a catalog of all stars that were visible (as faint as possible, and that was ultimately magnitude 18.5, so one hundred thousand times fainter than the naked eye can see), with accurate positions and apparent magnitudes (brightnesses).

A number of special problems arose. A giant telescope like the 60 in., which was the largest in the world at the time, had serious imperfections in imaging outside the center of the focal plane. That is because for a good image the mirror had to be parabolic, which only worked properly exactly in the center of the field being photographed. Just next to the 'optical axis' the images are of poorer quality and corrections are required to derive the magnitudes of the stars from the photographic plate. This turned a much bigger problem than

Fig. 9.18 Frederick Hanley Seares. This picture comes from the *Album Amicorum*, presented to H.G. van de Sande Bakhuyzen on the occasion of his retirement as professor of astronomy and director of the Sterrewacht Leiden in 1908 [47]

with the smaller refractor telescopes for example as used for the *CPD* and the like. It was mainly thanks to Frederick Seares that these corrections could be made. Furthermore, the brightness and positions had to be uniform and consistent within and among each of the *Selected Areas*. In order to measure bright stars consistently with the fainter ones, exposures had to be taken in which either the exposure time was shorter or part of the mirror was covered with an aperture. Seares, together with staff member Harlow Shapley, recorded special plates for this purpose between 1913 and 1919. Also, the magnitudes between different *Selected Areas* had to be consistent and related to a set of standard stars around the North Pole (which were visible all year round from Mount Wilson). This was done by recording both a *Selected Area* and this so-called *North Polar Sequence* (of standard stars) on the same plate. This was done with a smaller telescope of 10 in. aperture, because it was only necessary to record the brightest stars. This part of the work was done by Milton Lasell Humason (1891–1972).

A total of more than a thousand plates had been taken. The deep 60 in. plates were copied in Pasadena and then sent to Groningen. There Kapteyn had arranged for assistants and calculators to measure the plates. These were largely paid for by the Carnegie Institution. The support of the University of Groningen for Kapteyn and his pioneering work was again far from overwhelming, although at that time he was definitely their greatest and most famous scholar. Kapteyn did not live to see the completion of the work. Thanks to the unceasing efforts of Seares and Kapteyn's successor in Groningen Pieter van Rhijn, the *Mount Wilson Catalog of photographic magnitudes in Selected Areas 1–139* was published in 1930 as a special publication by the Carnegie Institution, eight years after Kapteyn's death. Of course, Kapteyn did see preliminary results.

An important consequence of Kapteyn's association with the Carnegie Institution and Mount Wilson Observatory was the introduction to America of young astronomers from the Netherlands and some other European countries. American astronomy was on the rise, observatories were built or expanded, and there was always a need for promising, talented young astronomers; so many positions, temporary or permanent, were available. Kapteyn opened a veritable 'pipeline', as the American astronomy historian David DeVorkin called it. Arnold Ernst Kohlschütter (1883–1969), Adriaan van Maanen, Ejnar Hertzsprung and Pieter van Rhijn, among others, profited from this.

Kohlschütter was a student of Karl Schwarzschild in Potsdam, with whom Kapteyn maintained good ties. He had obtained his PhD in Göttingen when Schwarzschild was still teaching there and now worked at the Bergedorfer Sternwarte near Hamburg. At Schwarzschild's request Kapteyn wrote a very strong recommendation to George Hale to arrange a visit from Kohlschütter. He left the same year and worked for a number of years with staff member Walter Adams. We will meet him again later.

Van Maanen was actually a student of Albertus Nijland in Utrecht. As we already saw, Adriaan van Maanen worked for some time in Groningen with Kapteyn and considered himself to be a student of Kapteyn rather than Nijland. His PhD thesis from 1911 was about proper motions of stars in the h and χ Persei clusters (see Fig. 7.10). Van Maanen was fortunate to have a rich cousin, who made it possible financially to volunteer to work at the Yerkes observatory for some time. He was personally delivered there by Kapteyn in 1911, when the latter was on his way to Mount Wilson. After that period Adriaan van Maanen went, no doubt with a strong recommendation from Kapteyn, to

Fig. 9.19 Van Maanen (far right) on part of a group picture of the staff of the Mount Wilson Observatory, apparently on the occasion of a visit by Hendrik Lorentz (second from the left). The picture is dated 'c. 1923'. The fifth from the left is Walter Adams, then director of Mount Wilson Observatory. Courtesy the Observatories of the Carnegie Institution for Science Collection, Huntington Library, San Marino, California [89]

Mount Wilson, where he eventually became a staff member and spent the rest of his career (see Fig. 9.19).

Ejnar Hertzsprung was originally a Danish chemical engineer, who was appointed by Schwarzschild in Potsdam in 1909 after some jobs at other places. He wanted to use the 60 in. of Mount Wilson and Schwarzschild suggested that he should contact Kapteyn. Hertzsprung came to Groningen and Kapteyn had arranged to accompany him to California in 1912. This was done, but by the time they left Groningen Ejnar Hertzsprung was engaged to Kapteyn's second daughter Henriette (Fig. 9.20).

Van Rhijn was a student in Groningen and was still working on a PhD thesis, which was completed in 1915. In his recommendation to Hale Kapteyn praised him as promising, but also wrote that 'whether he would become a good observer remains to be seen'. Van Rhijn went to Pasadena in 1912 and stayed there for some time. The correspondence between Hale and Kapteyn mentions in February 1913, that van Rhijn, while on Mount Wilson, had been informed of his father's death after Hale had received a telegram (probably from

Fig. 9.20 Ejnar Hertzsprung as a young man. Aarhus Universitet [90]

Kapteyn). He was very moved by the news and Hale advised him to take a short holiday.

For decades, in order to get a job as an astronomer at a Dutch university, it has been (and still is) required to work some time abroad. This then goes back to Kapteyn's time. In the meantime, 'postdoc' positions have become commonplace and if one aspires to a career in astronomy in the Netherlands, applying for those temporary positions is a first step after completing a PhD thesis. Some Dutch astronomers, including myself, followed in Kapteyn's footsteps almost literally and worked as fellow of the Carnegie Institution in Pasadena at what had become the Mount Wilson and Palomar Observatories. Many Dutch astronomers stayed abroad; some became directors of American observatories. There is an anecdote that when a young astronomer at a Harvard conference explained that he came from the Netherlands, Harlow Shapley remarked: 'Oh, that's where they grow tulips and astronomers for export!'

The annual trips to the United States also enabled Kapteyn to maintain his contacts with American astronomers. On his way to and from Mount Wilson, he regularly visited other observatories and astronomers who were friends of his. He also received various awards. Already in 1907 he was appointed as 'Foreign Associate' of the American Academy of Arts and Sciences. In April 1913 he gave a special lecture in Washington for the Academy on the structure of the Universe. He had to travel to America especially for this occasion—and be back in time for the wedding of his daughter Henriette to Ejnar Hertzsprung. In the same year he was awarded the very prestigious Catherine Wolfe Bruce Gold Medal of the Astronomical Society of the Pacific for his contributions to astronomy. He received this during his regular visit to California that year.

9.8   End of an Era

In January 1914 Kapteyn received a piece of very sad news: the death of his great friend David Gill after a short but serious illness. The Kapteyns traveled to London to assist the widow Gill as well as possible. Gill was buried in his hometown of Aberdeen, but a memorial service was held in London at the church of St Mary Abbots in Kensington. Kapteyn and many friends and staff attended that service. David Gill has been crucial to Kapteyn's work and the latter wrote a long in memoriam, which was published in *the Astrophysical Journal*, the leading American journal. Gill's death was the end of an era.

In 1914 the First World War broke out and that marked the end of another era: that of the annual visits to Mount Wilson. The Kapteyns happened to be there when the war began and they had great problems returning to Europe. It became a voyage in permanent danger because of German mines and submarines. With delay they reached the Netherlands. In a letter of January 23, 1915 to Frederick Seares Kapteyn wrote:

> Our journey has been without any difficulty. We were stopped by a man of war (English) when we were no more than a few hours out of New York. A couple of officers came on board, but they only examined the ship's papers. After that we came unmolested to Portsmouth, where, after some hours waiting, we got a pilot, who had to bring us safely from here up to near Harwich. We need him not so much on account of the mines, the situation of which (as far as they had not broken loose) was well known to our captain, but on account of the light houses which partly were kept dark, partly had been displaces on purpose.
>
> In Dover harbor we spent a night. The ship's papers had been taken to the shore, but even if they had not, our captain did not care to go by night. The permit to leave reached us somewhat late the next morning, so that we could not land at Rotterdam that same night, but had to lie 'on stream' off Hoek of Holland. Altogether we lost about 2 days, but that was all.
>
> Mrs. Kohlschütter had no trouble at all up to Rotterdam. We have not yet any news of her and cannot say therefore how she fared at the border.

Kohlschütter himself had left the United States shortly after the outbreak of war. He had tried to reach Germany via Gibraltar, but was arrested there and held prisoner of war for the rest of the war. Eventually he was transferred to England, where Arthur Eddington enabled him to do some astronomical work. The war prevented further visits to Mount Wilson. Kapteyn actually has never returned there. He did remain a paid Research Associate of the Carnegie Institution for the rest of his life, although he had great difficulty accepting this, so that Hale had to convince him a few times to do so, saying that the Institution appreciated his work for them enormously and did not care where it was done. It had been a very productive and auspicious time for both parties. George Hale later wrote to Willem de Sitter:

> Nothing has been so valuable to the Mount Wilson Observatory as the inspiration and guidance of Kapteyn. His splendid imagination and fine optimism have stimulated us to our best efforts and encouraged us to attack problems of wide scope.

Fig. 10.0 Kapteyn during his visit to Mount Wilson in 1908. The photograph was taken in the library of the 'Monastery', the guesthouse, where Kapteyn probably stayed and had a desk to work. Kapteyn Astronomical Institute, University of Groningen

10

Statistics and Other Concerns

I couldn't claim that I was smarter than sixty-five other guys
– but the average of sixty-five other guys, certainly!
Richard Phillips Feynman (1918–1988)

One has to be unselfish to a certain extent...
if only out of selfishness.
Marie von Ebner-Eschenbach (1830–1916)

After his last visit to Pasadena and Mount Wilson and the problematic return to Groningen in January 1915, Kapteyn was almost 64. Normally he would have more than six years to go before his retirement; this was in those days effected at the end of the academic year in which the person involved reached the age of seventy. His children had left the household already for some time. We already saw that the youngest, Gerrit Jacobus, was now a mining engineer and lived in Colorado. The second daughter, Henriette Mariette Augustine Albertine, was married to Ejnar Hertzsprung, who had returned to Potsdam after his stay on Mount Wilson and they had settled there together after their marriage. The war made contact between the Hertzsprungs and the Kapteyns more difficult, but since the Netherlands had remained neutral, this was still relatively easy. The eldest daughter, Jacoba Cornelia, had completed her medical studies and in 1906 had married Willem Cornelis Noordenbos, who had also studied

---

The quote is from Richard Feynman's *Surely you're joking, Mr. Feynman!*.
Marie von Ebner-Eschenbach was an Austrian writer.

© The Editor(s) (if applicable) and The Author(s), under exclusive license
to Springer Nature Switzerland AG 2021
P. C. van der Kruit, *Pioneer of Galactic Astronomy: A Biography of Jacobus C. Kapteyn*,
Springer Biographies, https://doi.org/10.1007/978-3-030-55423-1_10

Fig. 10.1 The Kapteyns in Groningen in an undated photo from a family album. Made available by great-grandson Jan Willem Noordenbos

medicine in Groningen. He became a lecturer in Utrecht (in 1920 he became a professor in Amsterdam) and they had four children between 1907 and 1915.

Once the children had left home, the Kapteyns moved in 1906 out of their comfortable upstairs apartment in the Heerestraat. In her biography Henriette Hertzsprung–Kapteyn [1] wrote (see also Figs. 10.1 and 10.2):

> In 1905 there had been a revolution in the Kapteyn household. When the eldest daughter was married, the second one was studying in Amsterdam and the son was still in Freiburg, the house became too big for them and they decided to move to some smaller quarters. Mrs. Kapteyn, who on her short visit to America had immediately developed a great sympathy for her American sisters and their regime of self-help, proposed to her husband, to break with the Dutch convention, to abolish the housekeeper and to take a lady for help in the household in the morning hours. She was the first in her circle to have this naughty idea, and her husband thought it was excellent, if the work that still was to be done would not be too much for her. A small upper apartment was rented on the Eemskanaal on the outskirts of the city, the housekeeper was dismissed and a maid was taken. Mrs. Kapteyn cooked herself and helped the maid. The people of Groningen did not think it was really *comme il faut*, but she enjoyed her work and the financial help she could give her husband. Why would they care about the opinion of others? In everything they were ahead of their time, for what is commonplace today was then unacceptable. And their lives always became fuller and happier. [...]

Fig. 10.2 The house at Ossenmarkt 6 in Groningen, as it appears today. The Kapteyns lived there on the first floor from 1910 to 1920. The entrance was in the narrow alley on the right side of the house. At the bottom left just above the hood of the left car is the plaque from Fig. 0.4. From Wikimedia Commons [91]

After five years they moved to a large old upstairs apartment at the Ossemarkt. This was a home 'with a soul'. People who visited it never forget it again. Not so beautiful, but very special was the entrance: a narrow alley like a crevice between two high houses and an insignificant door with an old-fashioned pulling bell, which was seldom used, because friends and acquaintances could open the door themselves, so that they could always enter freely. The beautiful, old oak staircase led to a hall, from which the large rooms could be entered with their atmosphere of serene tranquility and conviviality, which appealed to everyone. The spacious living room, with its old paneling, antique wallpaper and deep windowsills, had the view of the quiet, distinguished square through the high windows with the orthodox-Dutch half-closed curtains.

10.1   Tides

Kapteyn had a broad interest and did engage in other fields of science, sometimes—as in the case of tides below—with little success, sometimes—in the case of statistics, in the next section—to help others and then spend a disproportionate amount of time on it. We saw earlier that in his first years in Groningen he had, among other things, done research into periodicity in the weather and climate with the help of growth rings in trees.

Kapteyn gave a lecture on the for him unusual subject of *The theory of ebb and flow* on February 27, 1909 for the Mining Society in Delft. It is not clear why he was asked to speak on this subject. As far as I know he had not done any research in this field and although it was somewhat related to astronomy, it was still first and foremost a subject for someone in earth sciences. His son had studied mining engineering, but not in Delft. The lecture was published in the Society's *Yearbook*, in the form of notes by one F.T. Mesdag, probably the mining expert Ferdinand Taco Mesdag (1886–1961). These notes also contain a number of footnotes, which are obviously comments by Mesdag.

Kapteyn was completely wrong. The tides on Earth are a consequence of the attraction of the Moon and to a lesser extent the Sun. Suppose (which Kapteyn also did in his lecture) that the Earth is covered with a large ocean without continents and that it is equally deep everywhere. Then consider the point on this ocean exactly below the Moon, so where the Moon is directly above. The Moon exerts a certain attraction on the surface of the ocean. And also on the Earth's surface at the bottom. But the latter force is smaller because the Earth's surface is further away from the Moon and the ocean is more easily distorted by the Moon's gravitation. The result is that the surface of the ocean is pulled up a bit more than the bottom, and so there will be a bulge towards the Moon. The same goes for the opposite side of the Earth, but the effect is smaller there because that side is further away from the Moon. However, one must also take into account that there are centrifugal forces due to the rotation of the Earth and especially due to the motion of the Earth around the center of gravity on the Earth-Moon system. That point is outside, but not far from the Earth's surface, and the effect is greatest on the side turned away from the Moon. This causes a bulge on that side as well, which is comparable to the one directly under the Moon. So the ocean surface becomes a kind of ellipsoid, like the ball for rugby or American football, with the largest diameter away from and towards the Moon. In such simple considerations it is assumed initially, that the Moon is always above the equator. In reality, the plane of the Moon's orbit makes an angle of about 6° with the Earth's orbit around the Sun, and that angle makes another 23° with the equator. So the position on Earth at where the Moon is directly above, varies in latitude from about 30° south to 30° north. That is a considerable deviation, so one will have to include it in every theory of tides.

In this idealized case with a large ocean, the long diameter of the ocean's surface remains oriented towards the Moon and travels around the Earth. Now at a certain position in reality we do not see a wave passing by every twelve hours, because the Moon also moves around the Earth; that causes a cycle of twelve hours and 25 min. In practice it is even more complicated, because the influence of the Sun also makes itself felt, although that effect is smaller. Do Sun and Moon work together (when they are in the same direction as seen from the Earth), then we speak of spring tide; if the effect is minimal of neap tide.[1]

Kapteyn had a completely different theory. He took the simple case that the Moon orbits exactly above the equator. The attraction of the Moon can be understood as consisting of one component perpendicular to the Earth's surface and another one parallel to it. The latter is zero directly below the Moon and at maximum at those places where the Moon is on the horizon. At the equator, that component is along or directly opposite the rotation of the Earth. Water (not the Earth's surface itself, because it is solid) is accelerated or slowed down as a result, so that water flows away from the place where the velocity is greatest. And then you get a 'valley' in the surface of the ocean right under the Moon. Kapteyn summarized the result as: 'There is low tide under the Moon and high tide 90° from the Moon'. Exactly the opposite of what is actually the case! Mesdag seems to think the same and wrote in a footnote: 'So this conclusion differs from the widely accepted view'. How Kapteyn came up with his theory and why he did not present the usual explanation, which by the way goes back to Newton, Laplace and others, is a mystery.

If high tide is directly below the Moon, then one can—one would think—check that by comparing the times of high and low tide at a certain position on Earth with the times when the Moon is highest above the horizon. So it would be easy to verify Kapteyn's theory. Unfortunately, it is much more complicated in practice because of the presence of the continents. The largest water surface on Earth is the Pacific Ocean, and that is where the tides are strongest. This results in a tidal wave, which propagates around Cape of Good Hope underneath Africa and Cape Horn underneath South America into the Atlantic Ocean. The Dutch coast is reached via the Strait of Dover or around the British Isles. The wave takes a day or two and so the tide is very much behind

---

[1] For readers with a physics background: what counts is the *difference* of the attraction over a relatively small distance. Gravitation is inversely proportional to the square of the distance, but the tidal effect is then inversely proportional to the third power. This means a relatively large effect over a small distance. This third power is also why the influence of the Sun is less than that of the Moon.

compared to the change in the position of the Moon. For Scheveningen, the tide comes an hour or two earlier (and also more than ten hours later) than the passage of the Moon through the meridian, but that high tide is in fact a tidal wave that originated about two days earlier. Along the Dutch coast, this difference is also not the same everywhere as in Scheveningen. In Vlissingen it is less than an hour, in Den Helder it is six hours, and it is almost ten and a half hours in Eemshaven, Germany. It is not as easy as it seems to choose between the accepted and Kapteyn's explanation based on such a comparison. Nevertheless it remains incomprehensible to me how Kapteyn could have been so wrong.

Kapteyn, in his lecture, further treated cases where the Moon is not above the equator, and differences in the tides with geographical latitudes. Today a lecturer would have mentioned the fact that the Moon always keeps the same side facing the Earth due to tidal action, but Kapteyn did not. That the tidal wave slows down the rotation of the Earth with an increase in the duration of the day by 1 s per about 50,000 years, and that the Moon as a result moves away from us by 3–4 cm per year, was not known.

10.2   Statistics

Kapteyn has also been extensively involved in the field of statistics. There were in Groningen two demonstration models, which he had designed and were built, probably in the workshops of the biology department. Fig. 10.3 (left) shows the classical 'normal distribution', also called Gauss distribution. This is the probability distribution that occurs frequently in nature. From the funnel above, small bullets are dropped down through a system of small horizontal plates. Between every two plates the hole is large enough so that the balls can easily pass and in the row underneath the plates are arranged in such a way that each little plate is exactly straight underneath a hole. Chances are the same each time that a small bullet that falls on a pin will pass it on the left or on the right. The bullets are collected at the bottom and that results the well-known 'normal' probability distribution. Many things (such as the height of persons) follow such a distribution quite accurately.

The 'machine' was not Kapteyn's original idea. Sir Francis Galton (1822–1911) was the first to design such an instrument and had called it a 'quincunx'. The version in Fig. 10.3 still exists in Groningen.

Kapteyn was interested in the 'skew' distribution. For illustration he used the example of the distribution of the size of berries. It is quite possible that the diameters of the berries are distributed as the normal distribution, symmetrical

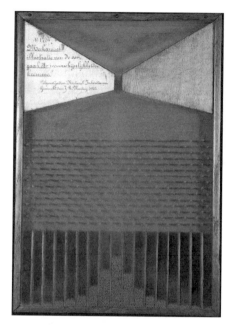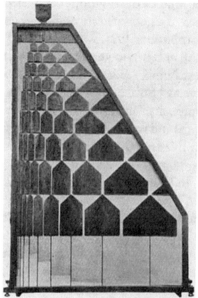

Fig. 10.3 Left: Demonstration 'machine' or quincunx for the normal or Gaussian distribution. This is a photograph of a still existing instrument in the biology department of the university of Groningen. It is believed that Kapteyn had designed it. For description see text. University museum Groningen. Right: Kapteyn's demonstration model for the skewed probability distribution. The original instrument is lost. This machine was designed by Kapteyn and built at the Groningen department of biology. Illustration from the publication of Kapteyn & Van Uven (1916), [92]. Such machines are called a quincunx

with respect to a certain value. But perhaps not the diameter but the volume is the relevant property. After all, volume is a measure of weight. Since volume is the third power of the diameter, you get a completely different, skew distribution. Suppose that the most common diameter is 5 mm and that the rest is symmetrically distributed around this with, for example, a quarter smaller than 3 mm, half between 3 and 7 mm and a quarter larger than 7 mm. Then the peak in the distribution of the volumes is at 16.5 mm^3, but the intervals of the quartiles are now at 8 and 101 mm^3. So very asymmetrical.

Kapteyn therefore had the quincunx built in Fig. 10.3 (right), which worked with sand instead of bullets. The holes through which the grains of sand fall are not equally spaced and, as can be seen at the bottom, we now get a completely different, skew, in this case 'log-normal' distribution. It is in fact a normal Gaussian distribution in which instead of plotting the horizontal axis on a linear scale now on a logarithmic scale. It is this kind of statistic that Kapteyn studied. In Appendix A.7 I give for aficionados a bit more mathematical background of the kind of distributions concerned. The instrument has been lost. However, there is a good picture of it in a publication by Kapteyn & Van Uven on skew distributions [92], which has been reproduced in Fig. 10.3 (right). Recently Florian L.R Lucas and Simon A.M. van der Salm, chemist and mathematician, have built a replica of the machine with financial support of the Royal Natural Sciences Society in which Kapteyn had been so active and donated it to the University Museum Groningen. Actually, the Museum contributed equally and Lucas and van de Salm put in money out of their own pocket. Their publication on this contains a thorough treatment of its workings and properties of the log-normal distribution [93].

In 1903 Kapteyn gave a lecture for the biologists at the University of Groningen with the title *Skewed frequency curves in biology and statistics*. The lecture was actually given in English, at least the text of the published version was in English. He began to say that he had been 'asked by several students and some other persons to discuss the statistical methods of Quetelet, Galton and Pearson'. The Belgian Lambert Adolphe Jacques Quetelet (1796–1874) and the British Francis Galton, the designer of the quincunx, and especially Karl Pearson (1857–1936) had developed and applied statistical methods for probability distribution and correlation between measurement in social and other sciences. In Groningen Jan Willem Moll (1851–1933) was professor of botany since 1890 and Jantina ('Tine') Tammes (1871–1947) was his assistant since 1897 (see Fig. 10.4). Jantina Tammes would become the first female professor of genetics in 1919. Moll had studied in Amsterdam with the famous biologist Hugo de Vries (1848–1935), who was very impressed by the use of statistics. De Vries and Kapteyn were to meet each other in 1904 in St. Louis as the only two Dutch participants at the Louisiana Purchase Exposition.

Kapteyn fulfilled the wish of the biologists, although he complained to others (e.g. in letters to Gill) that it took him too much time. Eventually he published the lecture. He described three applications. The first concerns data from his friend, the psychologist Gerardus Heymans, and dealt with thresholds in people's sensations. The second was from a publication by Pearson on house prices in England and Wales, and the third was a data-set on the size of crabs' foreheads in Naples. In the end he showed that these distributions followed the log-normal very well.

The article, which as mentioned was written in English, was apparently also read outside the Netherlands, because in 1905 Pearson wrote a reaction to Kapteyn's work in very harsh terms. In the first place, Kapteyn had not referred to the work of Francis Ysidro Edgeworth (1845–1926), who had developed a comparable statistic method as Kapteyn. Furthermore, he criticized a number of mathematical matters and especially Kapteyn's proposition that the form of a distribution function should contain causal information about its origin.

Fig. 10.4 Jan Willem Moll and Jantina Tammes were biologists in Groningen. University Museum Groningen, [94, 95]

He also claimed that the generalization proposed by Kapteyn had already been introduced by Gustav Theodor Fechner (1801–1887) and himself.

The tone was remarkable and unnecessarily harsh, so Kapteyn felt he had to respond to it. He admitted that he did not know Edgeworth's work and apologized for that. But he insisted that he should get at least part of the credit for the equations presented in the lecture, because Edgeworth's remarks had not yet been developed into a general theory. He also accused Pearson of having adopted his theory and general formula without giving him the credit for it. He also ridiculed Pearson's work to a certain extent by calling it nothing more than empirical. Pearson was furious and wrote a reaction to that as well. He particularly addressed the question of who was the first to propose the equations, and claimed that he had been using them for five or six years in his lectures on statistics, before Kapteyn came up with them. In short, he should have all the credit.

Kapteyn and Pearson were both apparently easily offended. It is plausible that they (and Edgeworth) did similar work with similar results independently of each other. But things never worked out between these two. In her biography Kapteyn's daughter writes:

> Typical of Kapteyn was his strong sense of scientific probity [integrity]. When in 1907 Frederick Cook had claimed to have reached the North Pole and later lied in his scientific report of the ascent of Mount McKinley, many people wondered

about the gullibility with which the scientific world had received those stories and had Cook's story assumed to be true. Kapteyn rose up against this with force; he considered it natural that *a priori* a scientific man should be believed at his word and not even to contemplate the possibility of misconduct. Tampering aroused his deep indignation. Even after years his wrath could flare up over a physicist who had come up with a theory, but when Kapteyn approached him to consult certain books to test the theory, he replied: 'But you cannot expect me to gather facts against my own theory!' And speaking of a well-known English biologist, in whose writings he had found evidence of tampering, he said: 'That's the one man I hate.' Twice in his life he personally encountered men who were not fair in science. He was horrified about this, it was impossible for him to put himself in that mental attitude. He was finished with him forever, they had failed in the elementary requirements of the *code d'honneur* and were, according to him, no longer worthy to carry the torch.

The English biologist in this passage was most likely Pearson.

Kapteyn returned to these matters in 1916. He said that he had decided to do this only after his retirement, because it took him too much time. But he was offered help by the mathematician Marie Johan van Uven (1878–1959), who had been a teacher at a gymnasium in Utrecht for a long time and had been appointed professor in Wageningen in 1913. The Agricultural University of Wageningen was not formally a university and actually had no professors, but they made an exception for him, because the Technical University of Delft (which had professors) had already offered him a professorship. Kapteyn and van Uven wrote a long article, which actually consisted of three articles (two by Kapteyn and Van Uven separately, on theory, and a third together on applications), [92]. In addition to mathematical details, they paid a lot of attention to Kapteyn's point of view that the form of a correlation should say something about its cause, i.e. why particular correlation should give rise to a certain form of the correlation function. This is no longer believed to be the case. An extensive study of the Kapteyn versus Pearson case has been published by Ida Stamhuis and Eugene Seneta [96], to which I refer for further details.

In the meantime Kapteyn had tackled another statistical subject, which he published in the leading British astronomical journal *Monthly Notices of the Royal Astronomical Society*. This occurred in 1912 and concerned the definition of the correlation coefficient. This is a number that indicates the extent to which two quantities, for example the mass and the luminosity of a star, are correlated. This coefficient is a number between zero and one, such that when there is no relationship between those two quantities it is equal to zero and unity if they correlate perfectly. Kapteyn wondered why one should use that number as such and not the square of it or maybe even a higher power or a root.

The coefficient is usually denoted by r. Suppose that a result is $r = 0.9$; how should you interpret that? If you take the square then you get $r = 0.81$ and that seems further away from unity. But if you take the square root then you get $r = 0.95$; that seems to be a much better correlation. So, Kapteyn wanted to know, what is the 'most natural measure of correlation'? Quite formal and subjective. It hardly seems like an interesting question, but Kapteyn had the opinion, which we now do not share with him anymore, that there should be something of a causal relationship in the correlation. In the end Kapteyn did come to the conclusion that it is best to simply take r. The definition of r in mathematical terms has been given in Appendix A.7.

But there was a catch. The quantity r was introduced by Pearson! Although it was based on Galton's work, the quantity is officially called the 'Pearson product moment correlation coefficient'. Now it is a fact that Kapteyn did not even mention the name Pearson! He did refer to Auguste Bravais, whom we met earlier in his study of the Apex motion of the Sun, and treated him as the founder of the study of correlation. It is not known if Pearson has seen this article by Kapteyn; as mentioned, it was published in a British, albeit astronomical, journal.

Kapteyn wouldn't give up easily in this kind of disagreement about priority and honor. In 1906, he approached David Gill and ask him to mediate to get an article on this matter in the renowned journal *Philosophical Transactions of the Royal Society*. Kapteyn believed that 'biologists under the influence of Karl Pearson were on the wrong track' with their study of skewed distributions. Even better had been the journal *Biometrika*, but Pearson was its general editor. David Gill wrote that he was prepared to do what Kapteyn asked, but in the end Kapteyn apparently did never manage to actually write such an article.

Kapteyn wrote to David Gill in 1907 that due to lack of time he would not be coming to Leicester, where David Gill, as President of the British Association, would give a lecture in which he wanted to speak prominently about Kapteyn, and for which he had been informed extensively about his work (see Sect. 6.5). Kapteyn mentioned in a letter that his involvement in statistical matters had got quite out of hand and wondered what kind of evil spirit had prompted him to do so. David Gill reacted by emphasizing that a gifted man like Kapteyn should not be concerned with things that others could look after equally well.

> Finally, what is the value or importance of a frequency curve in relation to the structure of the Universe? I am happy that you admit there is an evil spirit involved. Some form of exorcism is necessary and I would like to apply it.

In spite of this, Kapteyn kept himself busy with statistics, probably partly in a sincere attempt to help his biological friends, but his stubborn desire to be proven right in his fight with Pearson has probably also played a role. Some form of recognition he did get. A review article in 2001 in a British journal about the history and background of skewed (log-normal) distribution by three Swiss scientists in biology, statistics and mathematics [97] concludes with the words: 'Because of the fundamental and comprehensive contributions to the understanding of skewed distributions some 100 years ago, we dedicate our article to the Dutch astronomer Jacobus Cornelius Kapteyn'. Pearson is not mentioned in that article.

10.3 Groningen University

When Kapteyn was appointed in Groningen, the university had a monumental Academy building (see Fig. 4.5), but this was largely destroyed in a major fire in 1906. However, it did not take long (until 1909) until a new Academy Building (Fig. 10.5) was opened.

The tercentennial of the university in 1914 was celebrated with sobriety. A highlight was a meeting in the Martini Church, to which the professors dressed in academic gowns walked in procession (see Fig. 10.6). Kapteyn is walking in the second row, a confirmation of his prominent role within the university. At that time there were only 38 'ordinary' professors, so the parade in the figure shows more than a quarter of them. Four of the thirty-eight taught

Fig. 10.5 The new Academy Building in Groningen, opened in 1909. Beeldbank Groningen [98]

Fig. 10.6 Procession of professors on the Grote Markt, the central square of Groningen, on their way from the Academy Building to the Martini Church on the occasion of the three-hundredth anniversary of the University of Groningen in 1914. Kapteyn is the third from the right. Groningen University Museum

theology, seven law, eight medicine, nine mathematics and natural science and ten philosophy and literature. The university had no more than 450 students; those of Amsterdam, Leiden and Utrecht each were more than twice as large. Only the Free University of Amsterdam was smaller with 50 students (the other current Dutch universities were founded after 1914). Of those 450 students, about 50 were studying mathematics or natural science.

Kapteyn had been appointed foreign member of the United States National Academy of Sciences in 1907, and in 1914 he had been nominated by George Hale as the formal representative of the American Academy of Sciences to the Groningen celebrations. There was an official letter pointing out among others that

> The National Academy of Sciences, rejoicing in the completion of three centuries of educational progress and scholarly research by the University of Groningen, sends greetings and congratulations through its distinguished Foreign Associate, Jacobus Cornelius Kapteyn. In searching for the secret of Holland's great achievements in science we may best look to the example set by this University. Removed from the haste and confusion of the world, and steadfast in their devotion to the search for truth, her scholars have quietly pursued their way for centuries. Thus have they built up a center of learning whose foremost spirits inspire and guide the men of science of all nations.

Fig. 10.7 The auditorium of the Academy building (Fig. 10.5) has a set of stained glass windows on the first floor. It features various scientists with on the right as seen from the inside physician and anatomist Petrus Camper and on the right Kapteyn. This window is on the left above the balcony in Fig. 10.5. Photograph by the author

In the period 1930–1953, in the auditorium of the Academy building a number of stained glass windows has been installed with depictions of prominent scientists from Groningen. Looking from the auditorium at the lectern, the window most to the front has the physician and anatomist Petrus Camper (1722–1789) and Kapteyn (Fig. 10.7). Kapteyn is holding a Jacob's staff, an instrument used in pre-telescopic days to measure the angles on the sky between stars or other objects.

10.4   De Sitter and Leiden

Kapteyn's institute had a minimal staff. The permanent support staff consisted of Teunis Willem de Vries, who was first described as a 'clerk' and later as 'amanuensis' (laboratory assistant), and next to him since 1901 one scientific assistant, which initially was Willem de Sitter. There would have been personnel from the university to provide for cleaning and maintenance of the laboratory, but support in education and research could only be financed from temporary funding.

De Sitter, who was born in 1872, would normally have to wait for a professorship until 1921 (he would then be 49 years of age), the year in which Kapteyn would retire. The chances of a second professor of astronomy in Groningen were in fact zero. De Sitter's contributions were considerable, according to the articles in the series *Publications of the Astronomical Laboratory at Groningen*. Some concerned research fitting in with Kapteyn's large program, but some concerned the field of his own PhD thesis research, namely the properties of the orbits and masses of Jupiter's four large satellites. He certainly qualified for a position where he could set up his own line of research. But the only two other Dutch professorships in the Netherlands were in Leiden, which was held by Hendricus van de Sande Bakhuyzen and in Utrecht by Albertus Antonie Nijland (see Fig. 10.8).

Nijland had studied in Utrecht and had defended a PhD thesis there twice, first in 1896 with Willem Kapteyn on a pure mathematics subject and in 1897 with Jean Oudemans on measurements, partly photographic, of a star cluster. Jean Oudemans was to retire at the end of the academic year 1897–98 and a vacant position would result. It is not unlikely that Nijland quickly wrote his second dissertation, now in astronomy, in view of the upcoming vacancy. Not surprisingly, the professorship in Utrecht and the directorate of the observatory were first offered to Kapteyn. Kapteyn did not accept. At that time the *Cape Photographic Durchmusterung* was nearing completion and Kapteyn was busy trying to find a solution for the problem of the distribution of the stars in space by means of proper motions. In 1896 he had opened his own Astronomical Laboratory in Groningen, although still in temporary housing, and perhaps he no longer had any special ties with Utrecht, where he had after all he had studied at the university. Anyway, he turned down the offer. Utrecht then indeed appointed Albertus Nijland to the position.

Of course, the vacancy in Utrecht came far too early for Willem de Sitter. But in 1908 Hendricus van de Sande Bakhuyzen retired. This position, too, was first offered to Kapteyn. The letter from the Leiden Faculty, in which this offer was made, has been preserved in the Archives of the Sterrewacht; it was

Fig. 10.8 Albertus Antonie Nijland (1868–1936). This picture comes from the *Album Amicorum*, presented to H.G. van de Sande Bakhuyzen on the occasion of his retirement as professor of astronomy and director of the Sterrewacht Leiden in 1908 [47]

an extremely generous offer. The Faculty only inquired what the conditions would be to persuade Kapteyn to accept an appointment in Leiden. Leiden and the Sterrewacht were closer to his heart than Utrecht and it must have been an attractive offer, which Kapteyn nevertheless rejected again. Undoubtedly a role in this was played in this decision that he had just accepted George Hale's offer to come to Mount Wilson every year and the appointment by the Carnegie Institution. He was to travel to California for the first time that year and implement his *Plan of Selected Aries* that would be launched under his coordination with the new 60 in. telescope. Not a good time to take charge of an observatory.

It would not have been an easy job either. Leiden no longer had the prominent position it once had. The Sterrewacht did not participate in the *Carte du Ciel* (although there was now a photographic telescope). Astronomy had also changed worldwide. From determining the positions of objects in the sky as precisely as possible, astronomy had become more and more concerned with the physical properties and understanding of the cosmos. This development occurred in the first place in the United States, partly because of the construction of many new large telescopes. In Europe a lot of energy was spent on

Fig. 10.9 Antonie Pannekoek. This picture comes from the *Album Amicorum*, presented to H.G. van de Sande Bakhuyzen on the occasion of his retirement as professor of astronomy and director of the Sterrewacht Leiden in 1908 [47]

the *Carte du Ciel*, in which American observatories did not participate, and this happened under rigid regimes and guidelines. In Leiden, when Hendricus van de Sande Bakhuyzen took up his post as professor in 1872, he had chosen to pursue positional astronomy with vigor and not to become involved in emerging techniques such as photography and spectroscopy. Antonie Pannekoek (1873–1960) (see Fig. 10.9), who had obtained his PhD in Leiden in 1902, wrote in his memoirs [99] on conservatism at the Sterrewacht: 'In this environment, where everything lived in the tradition of twenty and thirty years earlier, where people were always performing calculations and nothing ever came of it, where the new paths of astronomy were scarcely appreciated, all the enthusiasm had to freeze out over time.' He describes how everyone gathered in the director's office on Monday morning, 'and I realized how after a week everything was still basically the same, only dribbled along a little bit. Then I always smelt the air of catacombs around me of deadly stiffness and boredom'.

Fig. 10.10 Ernst Frederik van de Sande Bakhuyzen. This picture comes from the *Album Amicorum*, presented to his older brother H.G. van de Sande Bakhuyzen on the occasion of his retirement as professor of astronomy and director of the Sterrewacht Leiden in 1908 [47]

On the other hand, Kapteyn saw that the future of astronomy in the Netherlands in the longer term had to lie in the first place in Leiden, where the Sterrewacht was fully supported by the university. At least in words—less in actual fact, when it came to maintenance and necessary extensions. But there were telescopes and at least some personnel to perform observations. In Groningen, the university's support was insufficient to realize major expansions. There was certainly potential in Utrecht, but Nijland also was not an inspiring astronomer. He observed variable stars, but did little in terms of interpretation or the application of new techniques. On the other hand, he was strongly focused on education and organized expeditions to observe total solar eclipses. Although the institutes in Groningen and Utrecht were not threatened with dissolution, Leiden seemed to have the most potential to play a leading role. Kapteyn must have realized that the Sterrewacht needed a strong leader, but also that a younger person (after all, he himself was 57 years of age in 1908) like Willem de Sitter would be better able to provide that leadership. Undoubtedly he insisted strongly that de Sitter would be appointed in Leiden.

Fig. 10.11 Willem de Sitter. The instrument at the telescope is a Zöllner photometer (see Sect. 2.3 and Fig. 2.6). This picture comes from the *Album Amicorum*, presented to H.G. van de Sande Bakhuyzen on the occasion of his retirement as professor of astronomy and director of the Sterrewacht Leiden in 1908 [47]

In Leiden the younger brother Ernst van de Sande Bakhuyzen was—as he already was in Kapteyn's Leiden days—still employed as an observator (see Fig. 10.10). For a while it seemed that de Sitter (Fig. 10.11) would be appointed as professor of astronomy as well as director of the Sterrewacht, a task that he, as a more theoretically oriented person, must have felt to be rather challenging. Eventually, however, de Sitter was appointed professor and Ernst van de Sande Bakhuyzen director of the Sterrewacht and extraordinary professor. The older brother Hendricus could not be denied some conservatism; the younger brother Ernst, however, was conservatism in the extreme. The story goes that he asked and got permission to continue using petroleum lamps, because he could not get used to electric light. De Sitter did not have his office at the

Sterrewacht, but was allocated a place to work in the director's house. He could make use of a number of calculators for his work, and was given every freedom to perform his own research plans. A thorough reorganization and modernization of the Sterrewacht Leiden was waiting to be implemented. As it would turn out, this had to wait another ten years or so, when Ernst van de Sande Bakhuyzen would turn seventy and go into retirement.

Fig. 11.0 Kapteyn on an undated photograph. University Museum Groningen

11

The Kapteyn Universe

One thing is sure. We have to do something.
We have to do the best we know how at the moment.
If it doesn't turn out right, we can modify it as we go along.
Franklin Delano Roosevelt (1882–1945).

In science it is better to be wrong than to be vague.
Often we find the right way only after we tried all the wrong ways first.
That is why it is fun to be a scientist.
You don't need to be afraid of being wrong.
Freeman John Dyson (1923–2020)

11.1 The First World War

The Great War, as the First World War is sometimes called, had in many ways an important, even decisive effect on Kapteyn, his work and his private life, as I am sure it must have had on that of almost everyone at the time. He had to give up his annual visits to Mount Wilson. But that was not such a big problem, as he had an excellent colleague in Frederick Seares, who very effectively kept the work with the 60 in. for the *Plan of Selected Areas* going. Dutch neutrality made it possible to continue correspondence, albeit with delays. Kapteyn, for example, maintained a lively correspondence with George Hale on a whole

_____

From Dyson's Oppenheimer Lecture (2000) [100].

P. C. van der Kruit, *Pioneer of Galactic Astronomy: A Biography of Jacobus C. Kapteyn*,
Springer Biographies, https://doi.org/10.1007/978-3-030-55423-1_11

range of subjects, as we will see below. The consequence of all this was also that Kapteyn never witnessed the completion of the 100 in. telescope, or saw the completed telescope. It was put into operation in 1917. In his letters to Hale, however, Kapteyn expressed the hope of coming to Pasadena and contributing to the telescope's scientific program. In the end it was too dangerous to make the crossing. The *Plan of Selected Areas* has never become a real subject of study with this 'Hooker telescope'. In 1919, Edwin Powell Hubble (1889–1953) was appointed at Mount Wilson Observatory, who found in the 1920s, among other things, that nebulae such as the Andromeda Nebula are systems like our Galaxy and that the Universe is expanding, which would become the permanent legacy of the 100 in.[1]

At the beginning of the war, Kapteyn demonstrated a remarkable neutrality. Just before the war broke out, in July 1914, he was informed that the German Kaiser wanted to honor him with the 'Orden pour le Mérite für Wissenschaften und Künste'. I use this mixture of German and French, because the website of the Orden uses this name.[2] This was a very special distinction which, although it had been awarded mostly to German scientists, only in special cases went to a foreigner, including Charles Robert Darwin (1809–1882), who received it in 1868, and physicist Lord Rayleigh (1903), as well as the Dutch Nobel Prize winners van 't Hoff (1901) and Lorentz (1905). Among the non-German astronomers who received the Orden were Simon Newcomb (1905), David Gill (1910), and Edward Pickering (1911). So it was a very great honor. According to the website of the Orden, Kapteyn was awarded it in 1915 together with the great physicist Max Karl Ernst Ludwig Planck (1858–1947). Whether he received the medal personally from the German Kaiser is not recorded, but it is likely. That the must have been in 1915 when Kapteyn visited Potsdam and saw his daughter and son-in-law; Potsdam is close to Berlin.

Although the honor was enormous, in August 1914 Kapteyn sent a letter to the German consul in the Netherlands, in which he stated that he would not accept the award because of the violation of Dutch neutrality by the German army, which had come to Kapteyn's attention as a rumor. However, the consul assured him by return mail that this information was completely incorrect and Kapteyn could do nothing else but change his mind. On the other hand, he never returned the decoration either. This issue of the 'Pour le Mérite' would emphatically get a follow-up after the war.

---

[1] In fact most of the redshifts were measured at Lowell Observatory by Vesto Slipher. For more background and discussion on the issue of Hubble's discoveries and his systematic failure to give the proper credit to others that they deserved, see the proceedings of a conference on *Origins of the Expanding Universe: 1912–1932* [101].

[2] Often in the literature the name is quoted with the French 'Ordre' at the beginning rather than the German 'Orden'.

Kapteyn spoke out strongly anti-German in letters to Frederick Seares and George Hale especially in relation to the important incident with the Lusitania. The RMS Lusitania was a British ocean steamer, carrying passengers. On May 7, 1915, the ship was torpedoed and sunk by a German submarine; more than half of the approximately two thousand passengers on board drowned. This incident played an important role in the United States' decision to give up neutrality and declare war on Germany. In a letter to Frederick Seares, Kapteyn called it an obvious example of 'barbarism', which had to open the eyes of those who still thought somewhat positive regarding the German case. 'That is not war, that is murder,' he wrote.

He must have been worried about his daughter Henriette. After all, she was married in 1913 to Ejnar Hertzsprung, who worked at the Astrophysikalisches Observatorium in Potsdam. As mentioned, the Kapteyns visited their daughter and son-in-law in Germany in 1915. Because of the predicament of the Hertzsprungs, which Kapteyn in a letter to George Hale shortly afterwards characterized with the German term 'unwohlwollende [uncomfortable or also malicious] neutralität', the war could not be a topic of discussion for them or for Kapteyn with many colleagues. He reported however that those with whom he actually was able to speak about it, had received great objections to this 'blödsinnige Krieg' (stupid war), which they had still felt was justified at the beginning. This was the intellectual class. Everything indicates that Kapteyn took the position that Germany was guilty of starting the war and that he supported the cause of the Allies.

Kapteyn was very concerned about the food supply and the welfare of his daughter and son-in-law. In 1916 Henriette gave birth to a daughter who was given the name Rigel—the brightest star of the constellation Orion. Rigel Hertzsprung (1916–1993) was born in Rotterdam, where Henriette stayed with her sister and brother-in-law Noordenbos, who both were physicians.

11.2 Radial Velocities

In addition to positions and luminosities of stars, spectroscopic observations were of great importance; after all, with a spectrum, the type of the star can be determined on the basis of the appearance and strength of dark absorption lines, which are characteristic of the atoms responsible for absorbing starlight in the outer parts. In the 1910s obtaining spectra was an enormously fast growing activity, which would eventually lead to an understanding of the structure, formation and evolution of stars. The important step of discovering the systematics of stars in the Hertzsprung–Russell diagram were a fundamental step, but for that the spectral type had to be known. But much more important for Kapteyn's program was that for a good understanding of the motions of the stars in space, the radial velocity had to be known. But one then had to be able to measure relatively small shifts of the absorption lines in the spectrum of a star, and to do that for many stars a large telescope was necessary. We have already seen that the Sun moves at 20 km/s relative to the nearby stars, so it is necessary to measure radial velocities with a accuracy of about 1 km/s.

The light of a star has to be dispersed over a larger area and this can only be done for the brighter stars to have sufficient signal on each part of the photographic plate that is used to record the spectrum. This kind of studies had already been done by Edwin Frost with the 40 in. refractor at Yerkes Observatory, the largest refracting telescope which had been built by George Hale before he started Mount Wilson, and by William Campbell with the 36 in. telescope at Lick Observatory. Especially the latter yielded good results, but only for the brightest stars, which meant in practice those visible to the naked eye. No wonder that the 60 in. telescope was extensively used for this type of measurement as soon as it was available. The person who did most of that work at Mount Wilson was Walter Adams.

Obtaining such radial velocities was part of Kapteyn's *Plan of Selected Areas*, but because it is only feasible for relatively bright stars this was not limited to the *Areas* themselves, but all stars of sufficient brightness were selected. In 1913, Campbell published radial velocities for 900 stars that were all brighter than magnitude 5, and in 1915, Adams came up with 500 stars between magnitude 5 and 8.

This has yielded important new results, particularly regarding the link between proper motion, radial velocity and spectral type. An important achievement was a publication by Kapteyn and Frost in 1910, building on earlier work by Frost and Adams, when Adams was still working under Frost at Yerkes. For a collection of 210 stars, in fact a sample of everything available in the literature about them, it appeared that if the velocities of these stars were corrected for that of the Sun towards the Apex, their average velocity depended strongly on the spectral type. This is, as we now understand, a consequence of their age. Stars are born of gas at low velocities relative to each other, but due to the gravitational force of the large complexes of gas from which the stars emerge, their average random velocity increases over time. It turns out that hot, bright stars are heavy and short-lived. So, these are generally very young, which is why they generally have lower random motions through space.

Kapteyn and Adams published a more precise analysis in 1915; undoubtedly they had worked together on this already before Kapteyn was prevented from visiting Mount Wilson. It was an analysis based on the velocities of Adams from Mount Wilson and by Campbell from Lick. This analysis confirmed the earlier result that the average radial velocity (in the line of sight) was smaller for stars with a smaller proper motion. Because small proper motion statistically indicates greater distance and thus intrinsically bright stars, this seemed to mean that the radial velocity and the absolute brightness (the intrinsic luminosity) of a star correlated with each other. That would say something about the properties of stars, but, much more importantly, in principle opened up the possibility of getting an idea of the distance of a star based on its radial velocity.

Adams investigated this further and found that for stars from which the distance could be determined independently with statistical methods, the radial velocity indeed was small if the distance was large and the star therefore intrinsically relatively bright. These were developments in which Kapteyn was very much interested. We will see later that this work would lead to serious friction between Kapteyn and Adams.

11.3   Correspondence with Hale

During the Great War, Kapteyn and George Hale maintained an extensive correspondence. Two themes stand out. The first is Kapteyn's appointment as paid Research Associate at the Carnegie Institution. Kapteyn felt, as we saw before, that he could contribute too little to justify this, especially now that he was no longer able to visit Mount Wilson due to the war. In a letter dated March 26, 1916, for example, Kapteyn stated that he hoped to return to California that year. This message crossed a letter from Hale dated March 9 (so mail did not travel quickly in those years, despite the neutrality of the Netherlands), in which he urged him not to come because of the risks. Kapteyn planned to come in mid-July, as in March his daughter Henriette was staying in Groningen because she was pregnant. And he remarked that the ship companies more and more often did not dare to offer tickets, so finding a ship for the crossing might even have been impossible. But then he concluded that if it would not be possible or desirable to come, he would resign as Associate of the Carnegie Institution. In fact, he also stated that in the five years that remained until his retirement, he had better concentrate anyhow on his work for his laboratory so as not to have to leave unfinished work behind. He was dissatisfied with what he has been able to do for Mount Wilson, and continuing the link with the Carnegie Institution was therefore undesirable. But in view of what George Hale has done for him, he wanted to consult him first before handing in his resignation.

Fig. 11.1 George Ellery Hale, painted by Seymour Thomas in 1929. This portrait hangs in the library of the Carnegie Observatories, Pasadena. Courtesy the Observatories of the Carnegie Institution for Science Collection, Huntington Library, San Marino, California [102]

George Hale (Fig. 11.1) answered on June 6 and was very outspoken. I quote part of that letter.

From what you say, to my great concern, about giving up your position as Research Associate of the Carnegie Institution, I am afraid I may have given you a false impression of the duties of Research Associates. It would be extremely desirable from the standpoint of the Observatory and the Carnegie Institution, for you to retain the position even if you never returned to Mount Wilson and never did a single piece of work for us. In other words, the purpose of the grant is to facilitate your own investigations, and not to secure any direct return to the Institution. This is done in the case of many other men, such as professor A.A. Noyes of Boston, and various others I could name if I had the Year Book here. So don't think of resigning under any circumstances. We want your close interest and counsel, but you need not do any work for us unless you have plenty of time for it.

On 23 July Kapteyn wrote back that he had renounced his intention to resign. But he would never return to Mount Wilson, not in 1916, nor in any other year.

The second theme concerns the conduct of scientific research. It started with a remark by Hale that it was pointless to make catalogs for the sole purpose of collecting data and not to gain more understanding. Kapteyn partly agreed with this. Catalogs are certainly useful, but the problem was that they were often made without any proper understanding of what they would be used for.

> The random way in which data have been collected in astronomy is astonishing. Take star positions, in a certain way the strong point because for over one hundred and fifty years such positions have been accumulating. Still, as soon as in stellar research you want particular positions you are pretty sure <u>not</u> to find what you want. So for instance data for the proper motions of stars fainter than 6th magnitude. So in many other instances. The trouble, I think, is that work was undertaken without having in view any particular problem for the solution of which the work is required. I know that many astronomers saw no other purpose in the colossal undertakings of the Astron. Gesellsch. Zone Catalog[3] and the Carte du Ciel than the providing of fixed points for eventual observations of comets and minor planets. Of course I knew that you hated 'catalogues' just on account of this, but I may have feared sometimes that your hate had led you too far. I am extremely glad to find that it is not so. My studies have made of me more and more of a statistician and for statistics we must have great masses of data of course.

Hale and Kapteyn agreed that the 'Art of Discovery' was being neglected too much. For Kapteyn this meant that people were insufficiently aware that they had to work through induction; many proceeded too quickly on to deduction and expected too much of it. 'Induction' was about taking stock of possible general conclusions from a limited amount of data or assumptions and selecting the most likely ones. 'Deduction' means drawing inevitable conclusions from general rules and laws. Induction is 'top-down': from observations or measurements one finds empirically an underlying law. Deduction is 'bottom-up': from a general principle or theory one derives specific consequences. It was said that deduction is more suited to mathematics and induction to experimental sciences, but these are generalizations.

Kapteyn took his Star Streams as an example. These are systematical motions of the stars. The deductive approach would have been: 'Take a wild guess, deduce the consequences and see if they match the observations'. How long

---

[3]The *AGK* was a German-led program to use meridian circles to determine the positions of stars up to magnitude 9 with even greater accuracy than before.

would he have had to go on before he would have come up with the idea of the Star Streams? His approach was looking for regular patterns in the motions across the sky, and from these he could immediately recognize—'il n'y avait qu'un pas', (it was just one step)—that there were two Streams. Hale admitted that induction played an important role in Kapteyn's discovery of the Star Streams. But he defended the view that 'a combination of deduction and induction offers more chance of success'. He clarified this through the process of making hypotheses, all of which have consequences and of which you can quickly eliminate several. The exchange of letters went on for a long time and the letters on this subject were detailed. Kapteyn rather stubbornly stuck to his view that the way of working had to be completely inductive. Hale insisted that he admitted that deduction certainly played a role as well. He suggested that Kapteyn should see Lorentz. Kapteyn first talked to his friend Heymans, philosopher and psychologist, as well as to Leiden theoretical physicist Paul Ehrenfest, Lorentz' successor. They both emphatically opted for Hale's more subtle and flexible point of view and not for Kapteyn's somewhat inflexible approach. But Kapteyn admitted no more to Hale than that between their points of view there could be no more than just a gradual difference.

11.4 Helium Stars

An important subject of Kapteyn's research concerned the so-called helium stars. He was already working on this in 1910, because he reported on the progress of this research during the congress of the Solar Union on Mount Wilson that year, when he gave a special presentation about the helium stars.

Helium stars are stars with particularly strong helium lines in their spectrum and correspond to the current spectral type B in the sequence on the horizontal axis of the Hertzsprung–Russell diagram. They are bright, heavy and short-lived stars. Such stars are also hot; the heaviest stars (spectral type O) are so hot that their surface the helium is ionized. Helium is the second element in the Periodic System and thus has two protons in its nucleus. In a neutral state, a helium atom has two electrons, but in the hottest O-stars, one has been detached. In the slightly less heavy and therefore slightly less hot B-stars that Kapteyn studied, helium atoms still have both electrons.

Of these stars, Kapteyn had found that their proper motions showed a high degree of what he called 'parallelism', at least unless you consider very large parts of the sky. This indicated that these stars would generally have a low velocity of their own in space relative to each other. This opened up a new possibility, because their proper motion would then be largely due to the velocity of the Sun and distances could be determined more accurately in the secular parallax approach.

Fig. 11.2 The proper motions of helium stars in the sky, drawn on Kapteyn's 'celestial spheres'. See text for details. From a 1914 publication by Kapteyn in the *Astrophysical Journal*. Kapteyn Astronomical Institute, University of Groningen

Figures 11.2 and 11.3 are from the first publication on this research in 1914. As a Research Associate of the Carnegie Institution Kapteyn published in the American journal *Astrophysical Journal*; this was via the Mount Wilson Observatory, where Frederick Seares was responsible for the organization's publications. In Fig. 11.2 we see Kapteyn's 'celestial spheres', which can also be seen on the picture of the Kapteyn Room in Fig. 0.2. These spheres were painted like a blackboard, so one could draw lines or write on it with chalk. The figure on the left shows how Kapteyn had divided the sky into a number of areas and had drawn the proper motions of individual helium stars. Indeed, they are quite parallel. In the figure on the right he had drawn the averages over those areas.

In the lower three areas, for example, the stars seem to move towards a common Apex. Combined with radial velocities one can in fact determine the individual distance of each star, just as was done for the Hyades (see Sect. 7.5). In this way Kapteyn derived the distance for each star individually with a reasonable accuracy.

In Fig. 11.3, which is from the same publication, Kapteyn has drawn in the positions of individual helium stars. The plane of the figure is that of the Milky Way and the stars are projected onto this. On the sky they are fairly close to the Milky Way, so in space they will also be near this plane. The figure is limited to an area visible from the southern hemisphere; this is so because the program of radial velocity measurements at Mount Wilson was not very far advanced yet. Eventually Kapteyn would have liked to repeat this work for the northern stars. The result in Fig. 11.3 shows how powerful the method was.

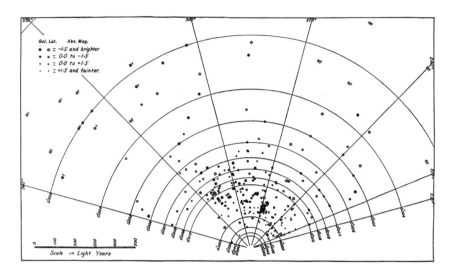

Fig. 11.3 The position of individual helium stars projected onto the plane of the Milky Way. The Sun is at the bottom where the diverging lines emanate from. The circles correspond to particular distances from the Sun. From a 1914 publication by Kapteyn in the *Astrophysical Journal*

Kapteyn's article contained over eighty pages, which was highly unusual at the time. But it was published in its entirety. In 1917, Kapteyn sent to Seares a second article that was even longer, this time about helium stars in the northern hemisphere. In it Kapteyn did something remarkable: in his analysis he used only the radial velocities of Campbell, determined by observations of Lick Observatory, and left out those of Mount Wilson. He quoted homogeneity as the underlying reason.

After concluding that for this type of stars individual distances can be determined, Kapteyn made a remark about stars of other spectral classes. He refers to work by 'Kohlschütter' or 'Kohlschütter–Adams', which seemed to point to the conclusion that one could deduce something about the luminosity of stars with spectra like the Sun from their characteristics; and this could then be used to determine distances. Nowadays this is a well understood effect. Among cooler stars there are giants and dwarfs, whose luminosity can differ enormously. The very bright giants are late stages of stellar evolution, while the faint dwarfs are stars with a lower mass than the Sun, which are on the Main Sequence. The spectra are very similar to each other and the temperatures at the surface are also similar, but due to the size and structure of the stars there is a large difference in the *width* of the spectral lines, which are much sharper and narrower in the giants than in the dwarfs.

11.5 Walter Adams

Kapteyn's collaboration with Walter Adams was severely damaged in the years of the First World War. In 1915, they had still collectively investigated the radial velocities of stars, based on the Mount Wilson program of measurements of such velocities. Now we have seen before that Kapteyn made an effort to arrange for promising young astronomers on Mount Wilson to spend some time there; not only his own students, like Pieter van Rhijn or Adriaan van Maanen, but he also mediated on behalf of Karl Schwarzschild from Potsdam, for young astronomers like Ejnar Hertzsprung. Another example was Arnold Kohlschütter, who came to work with Adams.

Adams saw Kapteyn's long manuscript about the northern helium stars and took offense at two issues in it. First, he felt he was being unfairly treated by Kapteyn's exclusive use of Lick radial velocities in this way, according to Adams, ignoring his observations. Of course Adams had a point here; after all, his telescope was bigger and with careful work he would obtain better results. Secondly, he was struck by the formulation 'Kohlschütter' or 'Kohlschütter–Adams' that Kapteyn had used, which gave Kohlschütter all the credit for discovering the relationship between the width of spectral lines and the intrinsic luminosity of stars, and downplayed the role of Adams. Adams wrote a letter to Kapteyn on this subject in June 1917 in which he complained in no uncertain terms. He reproached Kapteyn for being too accustomed to working with subordinates and therefore forgetting how to treat those who work independently. And Adams reminded him of how, when Kapteyn still visited Mount Wilson, he had repeatedly pointed out systematic effects, which he had only later put on more solid ground together with Kohlschütter.

Kapteyn was also sometimes rather hot-tempered and haughtily wrote to George Hale; he resigned again as Research Associate on Mount Wilson. In turn, he accused Adams of misconduct. The work that had appeared as Kapteyn & Adams in 1915, had been continued independently under the agreement to publish the result jointly. But despite this Adams had published his part under his own name. Kapteyn also mentioned to Hale how he personally had seen that Kohlschütter performed the work in question independently, while Adams now presented it as largely his own.

Through the mediation of Hale and especially Seares, the case was eventually settled. Frederick Seares rewrote the text in Kapteyn's article, which was published in three separate issues in 1918. Kapteyn was probably not quite correct in his evaluation of the situation. He indeed often reacted disproportionately strongly when he felt that the credit for a discovery was in danger of being wrongly taken from him or from someone else, and in this case he felt he should choose Kohlschütter's side. On the other hand, Adams was also a hothead, who has had similar problems with others. For example, he openly resisted the fact that so many European astronomers were given jobs in the United States which, according to him, actually belonged to Americans.

Almost immediately after the previous clash there was another run-in with Adams. After Kohlschütter had left California at the outbreak of the war, he continued his program of radial velocities with the originally Swedish astronomer Gustaf Strömberg (1882–1962). He was appointed at Mount Wilson in 1917 and continued the work with Adams on the relationship between radial velocity and absolute magnitude or luminosity. Strömberg wrote two manuscripts on this subject, one with himself as sole author and the other with Adams. Kapteyn did not object to the result itself, but he did object to the way in which average distances were estimated in these manuscripts. Kapteyn had derived a formula for this, which indicated how the average parallax of stars depended on the mean and apparent brightness (magnitude) and proper motion. Strömberg and Adams had adjusted his formula and Kapteyn was outraged that they dared to question his equation. George Hale responded to the matter by asking Kapteyn to write an extensive argumentation of the case, which he did; Frederick Seares reviewed the case as an independent referee. Seares solved the case again with great tact. The article of Strömberg by himself was published as part of the publications of Mount Wilson, but not—as was usual—first in a professional magazine which would sent it to anonymous experts for review. Strömberg admitted in his article that he had followed an incorrect procedure in the earlier manuscript with Adams. Seares ultimately blamed everything partly on Strömberg's poor English, but above all blamed himself for his own inattentiveness. Kapteyn and Adams would never work together again. Kapteyn is certainly to blame for his inflexible attitude; it is incorrect to put the blame for everything on Adams. But Adams was easily offended. After Kapteyn's death, his successor Pieter van Rhijn also got into trouble with Adams, who felt (unnecessarily, in my opinion) offended by certain passages in van Rhijn's publications.

Incidentally, Strömberg's work led to the discovery of an important concept called 'asymmetric drift'. The stars in the disk of the Galaxy move systematically around the center, the rotation of the Galaxy. If stars were to move in circular orbits, they would have to do so at a speed that produces a centrifugal force that would exactly compensate for the gravitational attraction of the matter in the Galaxy as a whole. But the stars also move relative to each other and do not move in circular orbits. The result is that for such an equilibrium between the forces, a lower collective velocity around the center will suffice than in the case of pure circular orbits. You can also look at it in terms of energy. There is a potential energy due to gravitation, and that needs to be compensated by kinetic energy in motion. If everything would move in circular orbits, then all kinetic energy would be in the collective velocity of the rotation, but if there are also random motions, part of the kinetic energy is contained in that, so less kinetic energy is needed in rotation. How much less depends then on how large these random motions are: the larger they are, the slower the systematic rotation is allowed to be.

If you now look at a group of stars that have average high velocities relative to each other, then they will as a collective move slower around the center of the Galaxy than a group in which stars move only little relative to each other. That first group will then be lagging behind in the rotation around the center of the Galaxy compared to the second group. This systematically slower motion through space of stars with large random velocities is called Strömberg asymmetric drift. This manifests itself, for example, in the case of young stars, such as Kapteyn's helium stars, which have small random velocities relative to each other. At the time of Kapteyn this was not yet understood; the asymmetric drift was only discovered and published in 1924 and 1925.

11.6 The Structure of the Universe

In the period after 1910, when the *Plan of Selected Areas* was well underway, Kapteyn published few papers for a while. Part of this was due to his work in the field of helium stars and statistics, which yielded few publications in the short term. It is also due to the fact that in Groningen he had to catch up with the lectures that he could not give during his visits to Mount Wilson. He still presented contributions to the Royal Academy and he also gave a number of general lectures for large audiences, which took a lot of time in preparation and execution. I will take a closer look at some of those lectures to follow Kapteyn's evolving insight into the structure of the Universe.

I have already mentioned his lecture on Mount Wilson during the great congress of 1910, and that to the American Academy of Arts and Sciences in Washington in 1913, for which he especially had to cross the ocean. Another important lecture took place when the organization Dutch Natural and Medical Sciences Congress (NNGC) held its meeting in Groningen in 1911. This NNGC, which was founded in 1887 and of which Kapteyn had become a member at an early age, had the aim of bringing together scientists from various disciplines; to this end it regularly organized congresses lasting several days. The NNGC still exists and now organizes a symposium twice a year on a specific subject. At the meeting in Groningen in 1911 Kapteyn was one of the keynote speakers.

On this occasion Kapteyn set out his ideas about the structure of the Universe. The title of his lecture was *A few recent researches in the area of evolution of the fixed stars and the Stellar System*. He began by explaining which different types of stars could be distinguished according to their spectra, and he expressed the suspicion that this would be an evolutionary sequence. According to him, the helium stars were then stars at a young, early stage and they would then evolve to a stage where they cooled down and would become similar to the Sun. That would continue until they finally became relatively red, cool stars (which we now know as M-stars). This is incorrect according to current insights, but as far as average ages are concerned, Kapteyn's view was not so bad. After all, the O- and B-stars (including the helium stars) are indeed heavy, bright and hot and burn their hydrogen very fast on the Main Sequence. So they are on average young, because they are short-lived. Stars like the Sun take much longer (the Sun takes about ten billion years), and such stars are on average older. The M-stars live even longer than the present age of the Universe, so they are on average the oldest. In terms of age, what Kapteyn said is true, except that it is not an evolutionary sequence at all, but a progression in how old stars can become.

Kapteyn also expressed the view that stars form out of the material (gas and dust) of which the nebulae and dark clouds also consist. That should then mean, that on average they have even smaller relative velocities than the helium stars. This also corresponds to our current insights. In earlier studies Kapteyn had extensively investigated whether nebulae had measurable proper motions, or whether their radial velocity could be measured. This was not possible in 1911, so it could not be verified. There were also nebulae that were called planetary nebulae, because they looked like a planet in a telescope with poor optics. A nice example, which Kapteyn also used to illustrate this, is the so-called Ring Nebula in the constellation Lyra (see Fig. 11.4 for an image obtained with the Hubble Space Telescope). Just like in that figure, there is often a star in the center of the nebula. Kapteyn now speculated that this might just be a late stage of a star, and that is correct. He probably came up with this because of his research of Nova Persei and the expanding nebula around it. He distinguished this from objects like the Orion Nebula (Fig. 9.10, in which stars are indeed forming) and the nebulae around the stars of the Pleiades (Fig. 9.12, which concerns reflection of starlight in dust and therefore *not* directly related to star formation). Kapteyn's ideas about the formation and evolution of stars are out of line with our current understanding, especially his interpretation that different spectral types are different stages of evolution, but apart from that he is not even very far off the mark in a general sense.

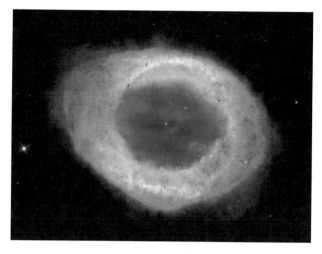

Fig. 11.4 The Ring Nebula in the constellation Lyra is an excellent example of a so-called planetary nebula. Picture from the Hubble Space Telescope [103]

It is also interesting that Kapteyn mentioned the spiral nebulae and even illustrated a few of them. He mentioned that there are 'hundreds of thousands, maybe even millions' of this kind of objects. But he did not address or speculate on what they were. Also in later publications he did not say a word about how

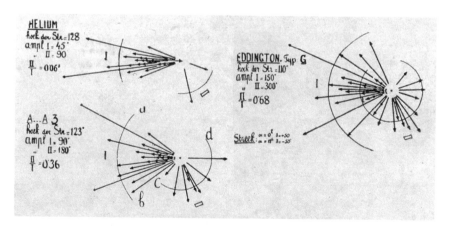

Fig. 11.5 The motions of the stars in the directions along the Milky Way plane for different spectral types. Top left the helium stars, which Kapteyn considered to be young. Bottom left stars of type A, which he saw as the next stage of evolution, and on the right the stars like the Sun, which would be the oldest. The direction to the left is that of Star Stream I. From Kapteyn's NNGC lecture, published in 1912. Kapteyn Astronomical Institute, University of Groningen

these spiral nebulae fit into the organization of the Universe or what place they occupied in it.

But then he picked up the subject of the Star Streams. For this he used Fig. 11.5 as an illustration. The three panels are more or less the three different stages of evolution of stars as he discussed it. Here we see the velocities of stars in projection on the plane of the Milky Way. Stars in the first Star Stream move to the left and it is clear that there is a big difference between these 'stages'. By the way, this kind of observations was reason for Kapteyn (and also for Arthur Eddington) to reject the interpretation by Karl Schwarzschild, that the streams in reality are no more than a reflection of asymmetry in the spatial velocities of the stars. After all, there should be no differences in the composition in terms of type of star. The inequality that Kapteyn stressed later disappeared with better observations, and Schwarzschild was eventually proven right.

Kapteyn thought a galaxy was formed from two moving parts, as shown in Fig. 11.6. It is not yet very clear to him exactly how this all worked in detail. The figure is schematic and he assumed that the directions of the velocities of the

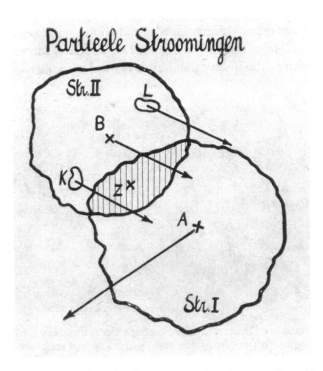

Fig. 11.6 The way Kapteyn thought about the two Star Streams. The 'z' in the overlap area indicates the position of the Sun. From Kapteyn's NNGC lecture published in 1912. Kapteyn Astronomical Institute, University of Groningen

two parts change over time. He also speculated about the role of gravitation and of an 'Ur-dust', out of which nebulae and stars are formed and which is not affected by gravitation. It is important to note that at this stage he already wondered what role gravitation plays in the structure and evolution of systems of stars. It is the beginning of what he would eventually present as the structure of the Galaxy, in which the distribution of the stars in space, their mutual attraction and their motions are such that the whole forms a consistent system that is in equilibrium. What we now call dynamics. But it was not that far yet.

11.7 The Spatial Distribution of the Stars

In the 1910s Kapteyn paid a lot of attention to how he would eventually, when sufficient data would be available, solve the problem of the structure of the Stellar System. The first thing to do was carrying the *Plan of Selected Areas* along far enough, but there were also separate investigations that addressed different aspects of the problem. Some of them were carried out by his PhD students. Their work has sometimes, but not always been published in the series *Publications of the Astronomical Laboratory at Groningen*, but in other cases it has remained a printed dissertation. One example is the study of proper motions of more than one hundred stars, on which Etine Imke Smid successfully defended a PhD thesis in 1914. As mentioned before, she was the first woman in the Netherlands to obtain a doctorate on an astronomical dissertation. After her defense, Etine Smid worked for Kamerlingh Onnes in Leiden for some time, but moved to Deventer after she got married and left science. Another example is the doctoral research from 1916 by Samuël Cornelis Meijering (1882–1943) on the motions of the cool K-stars. He found among these motions the same parameters for the two Star Streams as were known for the other stars. Gerrit Hendrik ten Bruggen Cate (1883–1963), who later changed his name to 'ten Bruggencate', got his PhD in 1920 on an investigation into the distribution of stars over spectral types in an area around the North Pole of the sky, using spectra that Frits Zernike had obtained in Potsdam. Meijering and ten Bruggencate both became teachers at the HBS.

But in the end it was all about the big picture: determining the distribution of the stars in space. We have seen that Kapteyn had to make three assumptions. The first of these was that stars move randomly in space, but that resulted in the Star Streams. He could re-check this by statistically determining the average parallax of groups of stars. The second assumption, of the absence of absorption of starlight by dust in space, had been thoroughly tested by him with the change of the color of the stars with distance. He actually found an effect that, according to current insights, is not so bad at all. However, Shapley's study of a globular cluster (see Sect. 9.5) had shown that there could be no so large an effect of reddening, so Kapteyn's result would be the result of systematic changes in color, for example as a function of the luminosity or absolute magnitude of the stars. Either way, absorption (or extinction) was believed by everyone in the astronomical community as being very small and could therefore safely be neglected.

The third assumption was that everywhere in space the stars are distributed in the same way over different sorts of stars, especially that the distribution of absolute magnitude or luminosity would be the same everywhere. But what that distribution, the luminosity curve as Kapteyn called it, looked like, still had to be determined. Pieter van Rhijn's dissertation consisted for half of a determination of the background light of faint stars, as described already. The second part was an accurate determination, with the latest data, of the average parallax of stars as a function of apparent magnitude and proper motion. If for a star one knew both properties, one could statistically estimate its distance. And to refine this, van Rhijn had determined this separately for stars of spectral class B (i.e., helium stars and the like) and for stars of types F, G, and K, i.e., stars roughly like the Sun (which is of type G). This then should lead to the determination of the luminosity function.

Of course, it was necessary to determine the star counts as accurately as possible and to the faintest brightness possible in the sky. By 'star count' is meant the number of stars as a function of their apparent magnitude. And that also, at least as a first exercise, as a function of Galactic latitude, i.e. of the angle above or below the plane of the Milky Way. Kapteyn had been working on this problem for much longer, first with Herman Weersma and, after he had left, with Pieter van Rhijn. A number of articles on this subject appeared in the *Publications of the Astronomical Laboratory at Groningen*, with finally in 1920 the, as far as Kapteyn was concerned, for that moment definitive tables of numbers of stars between certain limits of magnitude, proper motion, Galactic latitude and spectral type. This result was partly based on measurements of proper motions of large numbers of stars on plates recorded at Helsingfors, Cape of Good Hoop and Potsdam. This would then have to be expanded and improved as the work on the *Plan of Selected Areas* progressed.

And then there was the theoretical work. How do you get from all those data the distribution of the density (number of stars per unit volume)? That after all was the ultimate goal. This presents a mathematical problem that is rather difficult. I give a brief summary of the mathematics involved can be found in the online version of Appendix (A.8) for those interested. For this purpose, procedures were already developed by Schwarzschild in Potsdam and by (Ritter) Hugo Hans von Seeliger. The latter (see Fig. 11.7) we met earlier as one of the candidates for the position in Groningen to which in the end Kapteyn was appointed. A few years later von Seeliger had become professor in München and director of the Sternwarte there. His career was mainly devoted to the same problem that Kapteyn was working on, but his approach was much more mathematical. He and Kapteyn were very different characters who had little in common; although they must have known about each other's work, they communicated little. The mathematical approaches of Schwarzschild and von Seeliger were fundamentally different. Willem Johannes Adriaan Schouten (1893–1971) wrote a PhD thesis in which he compared the two methods. He was not a paragon of diplomacy; Schouten completely disagreed with von Seeliger's approach and presented that it in a rather antagonistic way. Kapteyn allowed the defense to go ahead in 1918, but gave the work of Schouten as little publicity as possible. Nevertheless, it seems to have had an unfavorable effect on the relationship between Kapteyn and von Seeliger. Schouten also became a teacher at an HBS.

Fig. 11.7 (Ritter) Hugo Hans von Seeliger. This picture comes from the *Album Amicorum*, presented to H.G. van de Sande Bakhuyzen on the occasion of his retirement as professor of astronomy and director of the Sterrewacht Leiden in 1908 [47]

For the sake of completeness I also mention Egbert Adriaan Kreiken (1896–1964) in the list of PhD students of Kapteyn. He had a special link with Kapteyn, because his parents had bought the *Benno* boarding school in Barneveld from Kapteyn's mother, after Kapteyn's father had died. So Kreiken and Kapteyn were born in the same house! Kreiken studied with Kapteyn and also started his doctoral research with his supervision; Kreiken always considered Kapteyn his real teacher, although he finally defended his thesis with Van Rhijn in 1923, when Kapteyn had already died. His research concerned the colors of stars in the Milky Way, based on photographic material obtained by Ejnar Hertzsprung at Mount Wilson. Kreiken continued to work in astronomy; eventually he founded the Ankara Observatory in Turkey, which is now named after him [104].

Fig. 11.8 Drawing of Kapteyn from 1920 by Cornelis Easton. This figure comes from an article on Kapteyn by the latter in *Hemel & Dampkring*

In 1920 Kapteyn (see Fig. 11.8) was finally ready to make a 'first attempt' to deduce the spatial distribution of stars. The article, written together with van Rhijn, was published via the Mount Wilson Observatory in the *Astrophysical Journal*. The first paragraph reads:

> The investigation contained in G.P. [Groningen Publications] 27, 29 and 30 were carried out for the purpose of making possible an elaborate treatment of the arrangement of the stars in space. At least two more publications will be necessary to complete this investigation. Now that, after so many years of preparation, our data seem at last to be sufficient for the purpose, we have been unable to restrain our curiosity and have resolved to carry through completely a small part of the work, even though, by doing so, the rules for strict economy of labor cannot be altogether adhered to.

Kapteyn and van Rhijn started by recalibrating the parallax of stars as a function of apparent magnitude and proper motion. In doing so they made use of the latest parallax measurements, both directly as a result of the motion of the Earth around the Sun and via the method of secular parallax, based on the velocity of the Sun through space. Then they had to make an assumption about how those parallaxes would be distributed around that average, but that—as Kapteyn had already shown—was not too critical. And from that they deduced what the luminosity curve was.

There's an interesting sideline. We already saw that the absolute magnitude, a measure for the luminosity of a star, is defined as the apparent magnitude if the distance would be 10 parsec. This definition comes from Kapteyn. He was one of the first researchers in this field and initially chose as the unit of distance that of a star with parallax of 0″.1. When Herbert Turner introduced the term parsec, Kapteyn first refused to use it, saying it was 'ugly'. In this article with Pieter van Rhijn, he finally accepted the term (although he kept calling it 'ugly'), but as a result felt compelled to redefine the absolute magnitude to a distance of 1 pc. At the first meeting of the International Astronomical Union in 1922 in Rome, the 10 pc was formally chosen for the definition of the absolute magnitude, taking over Kapteyn's decades of use (without mentioning Kapteyn by the way).

With methods that Kapteyn had already designed in principle in 1902, it was now possible to determine the luminosity curve using the available material. It resembled a Gaussian curve, as in the probability distribution, and that made the further mathematical procedures easier to perform. This luminosity curve is quite adequate for the bright stars compared to modern insights, but not so for the fainter stars. The curve of Kapteyn & van Rhijn rises for fainter absolute magnitude (the fainter intrinsically, the more stars there are in each volume of

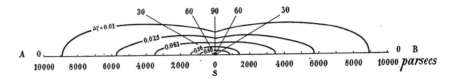

Fig. 11.9 The distribution of the stars in space according to Kapteyn & van Rhijn in their 1920 publication. The lines connect areas of equal density of stars in space. Kapteyn Astronomical Institute, University of Groningen

space) to stars about ten times fainter than the Sun. Nowadays we know that the curve from thereon flattens out, but in the article of Kapteyn & van Rhijn it goes down again. Because those are faint stars, it turns out that it does not make much difference to the star counts and thus to the final solution.

What is needed to determine the density of stars as a function of distance from the Sun is the counts of stars as a function of the apparent magnitude. After all, if you assume that the luminosity curve is the same everywhere, then those counts in the sky are only determined by the run of the density of stars with distance. It is not that simple to solve mathematically, but that is a problem that can be solved. As Schouten had discussed in his dissertation, there are two methods, and Kapteyn and van Rhijn opted for that of Karl Schwarzschild. They started using the counts in the Milky Way and an area around it, and averaged over all longitudes. In this way they found how the density changes when you move away from the Sun. That was a slow decrease, and by definition (because they averaged the counts) the Sun is then at the maximum of the density.

Next they took wide strips parallel to the Milky Way and averaged the counts on both sides. That way they found a decreasing density for 30° from the plane of the Milky Way, as well as for 60° and the direction of the poles, so the latter is perpendicular to the Milky Way. The result of all this they presented as a drawing, which is shown in Fig. 11.9. This is only the 'upper' half of the system; the lower half is symmetrical with respect to this. There also is symmetry between left and right as a result of the averaging of star counts. There was also some extrapolation, but the Galaxy they found had a size of 19,000 by 5,000 parsec. The outer contour corresponds to a density of stars of about 10% of that in the center. Because they averaged over Galactic longitude the Sun is by definition in the center.

11.8 Dynamics, the Crowning Achievement

The result that Kapteyn and van Rhijn published in 1920 was what Kapteyn had always been looking forward to. Around the same time von Seeliger had come to a fairly similar result, but he presented it much more schematically and in a much less accessible way. Kapteyn went further. What he actually wanted to know is not only how stars are distributed, but how the Galaxy could be in equilibrium, because there had to be a balance between the gravitational attraction of the stars on each other and their velocities. It is this next step that makes Kapteyn's work immortal; whether or not the model turned out to be correct in time, it is the crowning achievement of his efforts.

Nowadays we call this field dynamics; Kapteyn called it mechanics. The balance between velocities and gravitation has to result in a steady state, a stable equilibrium. He wrote about this several times in his letters to George Hale. In the correspondence about induction versus deduction, which I discussed above, there are also passages about this. I quote (from Kapteyn's letter of September 23, 1915):

> One of the somewhat startling consequences is, that we have to admit that our solar system must be in or near to the center of the Universe, or at least some local center. Twenty years ago this would have made me very skeptical... Now it is not so. [Von] Seeliger, Schwarzschild, Eddington and myself have found that the number of stars is greater near the Sun. I have sometimes felt uneasy in my mind about this result, because in its derivation the consideration of the scattering of light in space has been neglected. Still it appears more and more that the scattering must be too small, and also somewhat different in character from what would explain the change in apparent density. The change is therefore pretty surely real.
>
> Even more important than the central position of the Sun seems to me to be that our result for the first time shows the evidence of <u>force</u> in the great Sidereal System. A rough computation leads to the conclusion that at a distance corresponding with a parallax of $0''.02$ the stars are under the action of a force equal to the attraction of a central mass having 5 million times the Sun's mass.
>
> This number is very considerably higher than the number of stars we assumed up to the present to exist in a sphere of a radius corresponding to $\pi = 0''.02$. But these of course may be completely dark bodies, of which we know nothing.

Kapteyn wondered how the system can be in equilibrium. The result of his work on this was published after his retirement in the *Astrophysical Journal* in 1922. The article has the title *First attempt at a theory of the arrangement and motions in the Sidereal System*, and it is described in the summary as a first attempt at 'a general theory of the distribution of mass, forces and velocities' in the Galaxy.

So how did Kapteyn work out this theory? First of all, he had to be able to calculate the gravitational forces of all matter in the system at any location. To do this, he first reduced it to an elliptical distribution, see Fig. 11.10. There are mathematical formulas with which Kapteyn could now relatively easily calculate these forces at any location, except for a factor that corresponded to the average mass of the stars. But that factor had to be the same everywhere. He then started with the vertical direction. It is clear that there must be a certain average velocity to keep everything in equilibrium. If the stars have a lower velocity then that, then they cannot come up and down far enough from the plane to explain the observed thickness. Kapteyn concluded from measurements in the literature that the average velocity of stars in this direction is 10 km/s. He then calculated that there would be an equilibrium if each star on average has a mass 1.4–2.2 times that of the Sun. And it was believed on the basis of the orbits in binary stars that the components of such systems together are on average 1.6 times heavier than the Sun. That lead to the conclusion that Kapteyn's system would be in equilibrium in the vertical direction if most of the stars were binaries and Kapteyn felt that this was not unreasonable. Then there was no need for the presence of any unseen matter in the form the 'dark bodies'.

This is a remarkable result: the distribution of the stars and their velocities are precisely balanced. In 1932, Jan Hendrik Oort developed a model of this vertical distribution in more detail and in modern times we can do the same for other galaxies—work in which I myself have been involved. The question of a need to postulate unseen matter of unknown nature is still open. By the way, the vertical distribution that Kapteyn presented in his model in Fig. 11.10, and the average velocity in that direction, correspond remarkably well with

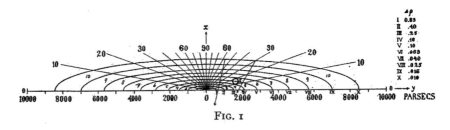

Fig. i

Fig. 11.10 The distribution of the stars in space according to Kapteyn approximated by ellipsoids. Compare Fig. 11.9, to which this should be an approximation

what we think this is now. His estimate of the total amount of matter that must be present in the solar neighborhood also is close to the best current determinations.

But what about in the 'horizontal' direction in the plane of the system? Again Kapteyn calculated the gravitational forces; he concluded that in order to have equilibrium and steady state there would have to be a centrifugal force, which would have to correspond to a rotation of the whole of the stars with a velocity of about 18–20 km/s. Now his Star Streams have a relative velocity of 39 km/s and so he could match everything if the Streams represented a rotation around the center of the system, one in one direction, the other one opposite to it. That meant then that the Sun had to partake in this rotation as part of one of the Streams and that it therefore could not be exactly in the center of the System. The modeling of the star counts had assumed symmetry as a first approximation so this constituted a refinement on that. After some considerations he came to the conclusion that the Sun is about 650 pc from the center; if it were less the centrifugal force would not be sufficient and if larger there would be a larger asymmetry in the star counts than observed. He also followed Ejnar Hertzsprung, who from studies of Cepheid variable stars came to the conclusion that the Sun is 38 pc from the symmetry plane of the Milky Way. These are part of the population of stars in the disk and their mean Galactic latitude shows a deviation from a straight line. From such studies of the shape of the Milky Way itself the best current value is 13 pc. Kapteyn indicated a position of the Sun in Fig. 11.10 with a small circle. This is indicative (as he himself stressed), but it is somewhat misleading; it is about a factor three too far from the center.

The fundamental theory of dynamics was developed elsewhere, especially in the U.K., by James Hopwood Jeans (1877–1946) and Arthur Eddington. It was Jeans who christened the model of Fig. 11.10 the 'Kapteyn Universe' when he studied its dynamics in more detail than Kapteyn's schematic treatment. Kapteyn was the first to present a consistent, dynamical model of our Stellar System, based on observations, in which the distribution of matter and the internal motions were precisely tuned to provide dynamical stability. It opened a whole new field of research and that is a major accomplishment.

Of course, it soon became clear that there is absorption of starlight and that the Star Streams do not represent a rotation around a center, but that does not detract from the originality of this brilliant application of a new approach to study the structure of the Stellar System and the dynamics of the Galaxy.

Fig. 12.0 Photograph of Kapteyn at the age of 70. This one comes from Easton's article *Personal Memories of J.C. Kapteyn* in the Dutch amateur periodical *Hemel & Dampkring*

12

Coda

Anyone who really delves into the lives of those
who have really accomplished something on this Earth
is always struck by the phenomenon that the actual strength
of those people has been a deficit turned into a quality.
Godfried Jan Arnold Bomans (1913–1971)

'Have we discovered our Galaxy yet?'
And I think the answer to this question is 'No, not quite'.
There is plenty of work ahead for the next generation of astronomers.
Heather Anita Couper (1949–2020) *[105]*

The speech that Heather Couper, astronomer and popularizer of profession, gave in 1985 as president of the British Astronomical Association, a society for amateur astronomers and astronomy for a wide audience, gave a nice overview of the history of our growing understanding of the structure of our Galaxy. If she is wrong somewhere, then it is, as so often, when it comes to Kapteyn's contribution to this understanding. American textbooks are often worse, because they describe Kapteyn (if they do mention him at all) as the one who was wrong about the structure of our Galaxy, while Shapley did present the correct picture. Heather Couper does not even do that: she only mentions Kapteyn as a pioneer of statistical astronomy and discoverer of the Star Streams; his role as a pioneer of international cooperation and as the founder of the study of

---

Godfried Bomans was a Dutch writer and television personality.
Heather Couper was a British astronomer, writer and documentary broadcaster.

© The Editor(s) (if applicable) and The Author(s), under exclusive license
to Springer Nature Switzerland AG 2021
P. C. van der Kruit, *Pioneer of Galactic Astronomy: A Biography of Jacobus C. Kapteyn,*
Springer Biographies, https://doi.org/10.1007/978-3-030-55423-1_12

dynamics remain unmentioned. The quote above is, of course, knocking on an open door.

12.1 After the Great War

For Kapteyn, the consequences of World War I were not limited to giving up his annual visits to the United States, California and Mount Wilson. International and national developments concerning the role of Germany after the war turned out to be a divisive issue, from which Kapteyn could not escape. The League of Nations was founded as a direct result of the Peace of Versailles, but Germany was not initially admitted as a member. This refusal to strive for reconciliation was not limited to politics; it played an increasingly important role in international science in general and astronomy in particular.

In October 1918, the British Society of London organized a meeting with exclusively representatives of the victorious countries (UK, USA, Italy, France, Belgium, Serbia and Brazil) on the creation of an over-arching organization of national scientific academies. Its report was published in an astronomical journal [106]. Kapteyn will certainly have read this. The tenor was that 'unlike after the end of earlier wars, personal relations between scientists of the allied countries and the central empires will be impossible for a long time to come'. The text of the report leaves no room for doubt:

> If today the delegates of the scientific academies of the allied nations and of the United States of America find it impossible to take up personal relations again, even in the matter of science, with the scientists of the central nations, inasmuch as these will not be admitted again into the union of civilized nations, they do this with full consciousness of their responsibility, [...]

> [...] the central powers have infringed the laws of civilization, disdaining all conventions and unchaining in the human soul the worst passions engendered by the ferocity of the struggle. War is inevitably full of cruelties [...]. These are not the acts we refer to, it is the organized horrors, encouraged and conceived from the beginning, with the sole aim of terrorizing inoffensive populations. The destruction of numberless homes, the violence and massacres on land and on sea, the torpedoing of hospital ships, the insults and tortures inflicted on prisoners of war, will leave in the history of the guilty nations a stain which the mere reparation of material damages will not be able to wash away. In order to restore confidence, without which all fruitful collaboration will be impossible, the central empires will have to repudiate the political methods the practice of which has engendered the atrocities which have roused the indignation of the civilized world.

In July 1919, the International Research Council was established, and Germany was excluded from membership. Countries that had remained neutral during the war, such as the Netherlands, were allowed to join at a later date, but care was taken to ensure that the statutes were fixed by then and could no longer be changed in case the neutrals still wanted to admit Germany.

Kapteyn fiercely resisted this, especially when it turned out that the Netherlands Royal Academy of Sciences was planning to join. His daughter wrote about this in her biography:

> Instead of reconciling, peace at Versailles encouraged irreconcilability, and long after the end of the war it glowed and boiled everywhere, both in the hearts of the conquered and in those of the conquerors. Science also continued to take sides, and in July 1919 the 'Interallied Association of Academies' was founded, to the exclusion of Germany.

> Kapteyn consulted with Heymans, what could be done to bring reconciliation in the scientific world. He had an unwavering faith in science and its eventual victory over bitter subjectivity. 'It is my conviction that science must in the long run directly and indirectly become a mighty factor in bringing peace and goodwill among men. If the men of science do give an example of hate and narrow-mindedness, who is going to lead the way?', he wrote in January 1917 to Eddington. In this great English astronomer, Quaker and strict scientific man, he saw a humane and impartial judge, who would not be blinded by hatred. He was an exception in that respect.

> When, at the end of 1919, an appeal was circulated to the neutral academies to join the International Research Council, but excluded Germany, he resisted with all his might, and tried to hold back the Amsterdam Academy of Sciences from taking this step. In the decisive meeting he and Heymans used all their influence to keep the members back, but they did not succeed. Here it was opportune to oppose idealism, and also in science idealism had to give way. He had not expected that and the shock was so great, that he and Heymans immediately left the Academy, which according to them had proven to be unjust and unable to act scientifically. It was more than a transient shock: until his death, this shaken confidence in justice and objectivity remained a painful wound.

Kapteyn and Heymans composed an open letter:

> To the Members of the Academies of the Allied Nations and of the United States of America' [107], in which they adjured that science should be 'the great conciliator and benefactor of mankind.' And they ended: 'We understand how your attention of late has been monopolized by what is temporal and transitory. But now you more than all others are called upon to find again the way to what is eternal. You possess the inclination for objective thought, the wide range of

Fig. 12.1 Hendrik Antoon Lorentz. This picture comes from the *Album Amicorum*, presented to H.G. van de Sande Bakhuyzen on the occasion of his retirement as professor of astronomy and director of the Sterrewacht Leiden in 1908 [47]

vision, the discretion, the habit of self-criticism. Of you we had expected the first step for the restoration of lacerated Europe. We call on you for cooperation in order to prevent Science from becoming divided, for the first time and for an indefinite period, into hostile political camps.

The physics department of the KNAW was chaired in that period by Lorentz (Fig. 12.1). These were turbulent times and the decision that the Netherlands would join will have resulted in deep differences of opinion among the members. Actually, Lorentz is praised by his contemporaries and historians for his wisdom and tact (see for more [108]). Kapteyn did not give up his membership of the KNAW as Henriette Hertzsprung–Kapteyn in her biography [1] seems to imply; he did, however, discontinue his attendance to the meetings, which must have been quite a sacrifice for an active member like him.

Kapteyn's standpoint in these matters also placed him in a difficult position in relation to colleagues at home and abroad, many of whom were great advocates of the exclusion of Germany, such as the British astronomer Herbert Turner of Oxford. But he also had supporters, such as Svante Strömgren

(1870–1947), director of the observatory in Copenhagen. The latter made an effort to provide Germans access to material that was denied to them, as in 1919 when others refused to share observations of a comet with German astronomers. 'Even my boys of 9 and 11 would never do such a thing', he said. Because of his influence, Kapteyn was asked to join the board of the Astronomische Gesellschaft. Kapteyn agreed, not because he aspired to such positions, but because it might help to repair the rift. He found Arthur Eddington at his side, who in 1920 was the only British astronomer who attended the congress of the Astronomische Gesellschaft. Eddington carried a lot of weight; after all, during the solar eclipse of 1919 he had verified the deflection of light from stars by the gravitational force of the Sun predicted by Einstein.

12.2 Reorganization of the Sterrewacht Leiden

We have seen that Willem de Sitter was appointed professor in Leiden in 1908, but that Ernst van de Sande Bakhuyzen became director of the Sterrewacht. I already described the conservatism of both the brothers van de Sande Bakhuyzen. On March 3, 1918 Ernst died suddenly, about half a year before his retirement. De Sitter became interim director and, with major involvement of Kapteyn, started a reorganization. This was an extremely important phase in Dutch astronomy, because now the foundations could be laid for the consolidation of the internationally prominent position that Dutch astronomy had assumed under Kapteyn, and which turned out to be a prelude to its spectacular success in the twentieth century. The focus of Dutch astronomy now was in Groningen, but Kapteyn must have realized that the conditions there were insufficient to maintain and strengthen this position; together with de Sitter he safeguarded the interests of Dutch astronomy by moving the center of gravity to Leiden. The curators of Leiden University supported its Sterrewacht and astronomy department wholeheartedly. Of course Kapteyn's Astronomical Laboratory in Groningen was maintained under his successor Pieter van Rhijn, but under his leadership it soon became less prominent. Under de Sitter the Sterrewacht Leiden became a prominent institution.

The ins and outs of this development have been described in detail by others, such as in Baneke's book *The discoverers of Heaven* [25] and elsewhere [109, 110]. The result was that Willem de Sitter became director of the Sterrewacht and headed a theoretical department. In addition, two other departments were created, under the leadership of two deputy directors. The astrophysics department was to be headed by Ejnar Hertzsprung, Kapteyn's son-in-law, who moved to Leiden from Potsdam for this and initially became an extraordinary professor. Kapteyn has always maintained that he did not give Hertzsprung any special advantage and as far as we know that he indeed did not. After all, Hertzsprung was a prominent astronomer already and did not need any special treatment. His appointment still took some time, but became effective in 1919, as did the appointment of de Sitter as director. The astrometry department would then be headed by Antonie Pannekoek. The latter presented a problem because of his strong socialist and communist views. The prime minister, Pieter Willem Adriaan Cort van der Linden (1846–1935), leader of the liberal party, who had held this position since 1913, had his doubts but was not prepared to oppose his appointment. However, after new elections a new prime minister appeared, Charles Joseph Marie Ruijs de Beerenbrouck (1873–1936), who was a catholic. He had different views than his predecessor and personally blocked the appointment of Pannekoek. In the end, Pannekoek was appointed in Amsterdam; the government in Den Haag had no say in selecting candidates for the municipal university and the Amsterdam city council did not share the objections.

For the time being, the gap was filled by appointing Kapteyn for one day a week after his retirement in 1921. However, he was only able to work in Leiden for a few months before he became fatally ill. The astrometric department did remain intact. With the Sterrewacht in the hands of Willem de Sitter, Ejnar Hertzsprung and not much later Jan Hendrik Oort, the future of both Leiden and Dutch astronomy was in very capable hands.

It is important to mention, however, that there were also developments in Utrecht, which in addition to those in Leiden, Groningen and Amsterdam were of great significance. Especially the appointment in 1918 of the Flemish Marcel Gilles Jozef Minnaert (1893–1970) was an enormous stimulus. Utrecht astronomy flourished until the unfortunate dissolution of the astronomy department in Utrecht in 2012, putting an end to a great tradition of 370 years (see a summary of events by the last chairman of the Utrecht Astronomy Department, Christoph U. Keller in [111]). The situation around 1920 is described excellently in a letter from de Sitter to Kapteyn dated May 24 of that year. At that time de Sitter was in Arosa, where he had to stay after being exposed to an overdose of ether during surgery. In his letter he discussed, among other things, the future of the Bosscha Observatory at Lembang on Java in the Dutch East-Indies. This Observatory was made possible by rich tea-planter and amateur astronomer Karel Albert Rudolf Bosscha (1865–1928) from Bandung and his nephew Rudolf Eduard Kerkhoven (1848–1918). This was formally opened in 1923, but planning had been started much earlier. Kapteyn felt it should be managed by a national committee, while de Sitter took the position that the Sterrewacht director (so himself) should be responsible for defining its research program and organizing the interface with Dutch astronomy. De Sitter argued that the Minister's attempts at specialization and concentration could lead to a division of tasks among Dutch universities. He repeated (so he said) his opinion regarding astronomy, which he summarized as: 'Leiden specializes in everything except the specializations of Groningen and Utrecht'.

12.3 Forty Years Professor and Seventieth Birthday

In 1918 Kapteyn had been a professor for forty years and several friends and colleagues wanted to celebrate that. Willem de Sitter and Pieter van Rhijn must have taken the initiative. Funding was raised and the famous painter Jan Pieter Veth (1864–1925) was commissioned to paint a portrait of Kapteyn. In the Kapteyn Room at the Kapteyn Astronomical Institute in Groningen—which contains much of Kapteyn's books, research materials, his 'celestial spheres' and the desk he worked at (see Fig. 0.2)—there is an album containing cards with the names of the persons who contributed financially to this (see Fig. 12.2). These are listed in alphabetical order, except on the first two pages (see the figure), on which Kapteyn's special friends and colleagues are collected. They are all persons who have been important in Kapteyn's career in astronomy or at the University of Groningen. The signatures in the two panels on the right in Fig. 12.2 are those on the first page of the album: Ursul Philip Boissevain, Gerard Heymans, Adolf Frederik Molengraaff, Jan Willem Moll, Pieter Johannes van Rhijn, Willem de Sitter, (Mrs.) Isobel Sarah Gill, née Black (1849–1919), George Ellery Hale, Edward Charles Pickering, Anders Severin Donner, Karl

Fig. 12.2 The first pages of the album that was presented together with a painting by Jan Veth on the occasion of Kapteyn's fortieth anniversary as a professor. From the Kapteyn Room at the Kapteyn Astronomical Institute, University of Groningen

Friedrich Küstner and Robert Thorburn Ayton Innes. The painting was presented exactly forty years after his inaugural address, on February 20, 1918.

In the album we also find a postcard, with sender 'Westersingel 19' with the text:

'Ad Jacobum Cornelium Kapteyn
Qui specula caruit, servare unde astra liceret,
sideribus fluctus vidit inesse duos.
a.d. X Kal. Mart. MCMXVIII'

The abbreviated last line reads in full: 'ante diem decimum Kalendas Martias 1918'. Experts on Latin texts and professors in Leiden Joan Booth and Herman Frederik Johannes (Manfred) Horstmanshoff, whom I consulted, translate the text as follows: 'He who lacked a telescope with which to observe the heavens saw that there were two streams of stars. 20 February 1918.' The date is in Roman notation '10 days before March 1, 1918'. According to the municipal register of the city of Groningen, Jacobus van Wageningen (1864–1932), then professor of Latin language and literature in Groningen, appears to be the sender. According to my informants this 'dactylic hexameter followed by a dactylic pentameter' is of very high quality Latin.

Ursul Boissevain, professor of ancient history and special friend of Kapteyn, presented the painting with a speech. The original text has been conserved in his neat handwriting in the Museum Boerhaave in Leiden. The closing paragraphs read:

And now, Kapteyn, accept this as it is brought to you, as a token of our cordial affection. May you be given many more years to come to explore in the unabated power of your more powerful mind and to discover, to penetrate ever deeper into the secrets of the immeasurable space of the infinite Universe, about which the mere sight already fills simple souls with the deepest respect.

Then the wish will be fulfilled, that we cherish for your dearly beloved wife, for your children and grandchildren, and for ourselves, for whom your friendship is among the most precious things we possess.

The painting presented to Kapteyn (and his wife) is now in the Kapteyn Room at the Kapteyn Astronomical Institute. It came into the possession of Kapteyn's eldest daughter Jacoba Cornelia Noordenbos-Kapteyn, who took it with her to England, where she lived close to her daughter Maria Newton-Noordenbos during the last few years as a widow. Now a coincidence was that this Maria was a classmate of Adriaan Blaauw, third director of the Kapteyn Astronomical Laboratory in the third grade at the gymnasium. After con-

Fig. 12.3 The painting by Kapteyn at his desk, which Jan Veth made on the occasion of his 40th jubilee as a professor. The person at the top right is David Gill. The painting is now in the Kapteyn Room of the Kapteyn Astronomical Institute in Groningen (see Fig. 0.2).

sultation with the other Kapteyn descendants, it was decided to donate the painting to the University of Groningen after Jacoba Cornelia's death (as well as the star chart in Fig. 2.2). The painting is reproduced in Fig. 12.3. Blaauw also mentioned the following about it in the proceedings of a 1999 symposium on Kapteyn, *The Legacy of J.C. Kapteyn* by myself and Klaas van Berkel [4]. Pieter van Rhijn, successor of Kapteyn and predecessor of Blaauw, had once told him that Mrs. Kapteyn did not like the portrait presented, because—she would have said—'I do not know him like that. I know him sitting at his desk.' Jan Veth has made a new painting and this now hangs in the Kapteyn Room (see Figs. 0.2 and 12.3). But what happened to the original painting? Adriaan Blaauw assumed that it was over-painted by Veth himself with a gown, jabot and beret, and that it was donated to the University to be placed in the gallery of professors in the Senate Room in the Academy Building when Kapteyn went into retirement in 1921 (Fig. 12.4). Blaauw examined that painting, together with the curator of the University Museum, and found traces of this

Fig. 12.4 The formal painting, also by Jan Veth, with Kapteyn dressed in academic gown, jabot and beret. It now resides in the Senate Room in the Academy building of the University of Groningen. University Museum Groningen

over-painting on it. At the presentation of my scientific biography *Jacobus Cornelius Kapteyn; Born investigator of the Heavens* in January 2015, Jack Kapteyn (Jacobus Cornelius), grandson of Kapteyn's son Gerrit Jacobus, was present. He had a painting with him that he possessed, and that also was painted by Jan Veth (see Fig. 12.5). This is clearly a preliminary study. Completely in accordance with Blaauw's hypothesis, the facial expression and posture are the same as in the painting with the academic gown; Kapteyn is wearing the same clothes there as behind his desk in the other painting.

As Kapteyn's seventieth birthday (January 1921) approached, Willem de Sitter took the initiative to publish a collection of Kapteyn's most important papers. He formed a committee, including George Hale and the British Astronomer Royal Frank Dyson. The latter received the support of Arthur Eddington, but also met with some very strong opposition in the United Kingdom. This was partly because the German Karl Küstner was also a member of the committee. Herbert Turner also was very much opposed to the plan;

Fig. 12.5 Preliminary study for the portrait that Jan Veth painted on the occasion of Kapteyn's forty years jubilee as professor. Made available by Kapteyn's great-grandson J.C. (Jack) Kapteyn

he especially blamed Kapteyn for accepting the 'Orden pour le Mérite für Wissenschaften und Künste" (see Sect. 11.1) of the German Kaiser. Others who had been on good terms with Kapteyn before, such as Arthur Hinks, joined the opponents. The latter accused Kapteyn of having received that award at the same time as the captain of the submarine which had sunk the Lusitania. This events had aroused a great deal of indignation, as we have seen, in fact also in Kapteyn, but that accusation turns out to be pure nonsense. The sinking of the Lusitania took place in May 1915 and the captain in question, Walther Schwieger (1885–1917), only received the medal (in the military class, by the way, not the scientific one) in July 1917 [112]. Kapteyn received his medal in 1915 (but had been informed of it in 1914).

Because of all this nothing came of the plan and Kapteyn's seventieth birthday passed relatively unnoticed. It was not a festive period in all respects anyway, because Kapteyn's 14-year-old granddaughter Greta Noordenbos had to undergo a major surgery in the United States after a meningitis. It ended well, but this granddaughter died a few years later. However, it was gratifying news that not long afterwards the government approved the naming of Kapteyn's laboratory, so that it formally became the Astronomical Laboratory 'Kapteyn'.

12.4 Kapteyn and Shapley

Kapteyn published his 'Kapteyn Universe' (Fig. 12.6) in 1922. Before that, however, there was already a discussion going on about the size of the Milky Way as a stellar system and its relation to the spiral nebulae, such as the Andromeda Nebula and NGC 891 in Fig. 6.7. In April 1920, the American Academy held a debate on the issue of *The scale of the Universe* between Harlow Shapley and Heber Doust Curtis (1872–1942), which came into the history book as the 'Great Debate'. Shapley defended the view that the Galaxy was large and that nebulae were part of it, while Curtis thought that the Milky Way was a smaller galaxy (comparable in size to the Kapteyn Universe published not much later) and that the spiral galaxies had similar internal structures. As we now know, Shapley was right about the first, Curtis was right about the second.

Shapley based his position on his determinations of the distances of globular clusters. As we saw earlier, these are spherical clusters of old stars in a more or less spherical volume (Sect. 6.2 and see Fig. 9.15); Shapley had determined from one of these globular clusters (M13) that the interstellar absorption by dust was negligible. In 1912 Henrietta Swan Leavitt (1868–1921) had discovered the existence of a class of variable stars, called Cepheids, whose period in a regular variation of brightness was a measure of its intrinsic luminosity. One can use that then to determine the distance. The variation is caused by the fact that these are somewhat unstable stars are in a late stage of evolution, pulsating periodically. Hertzsprung was the first to apply Leavitt's discovery in

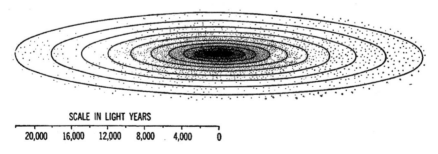

Fig. 12.6 The Kapteyn Universe. Kapteyn Astronomical Institute, University of Groningen

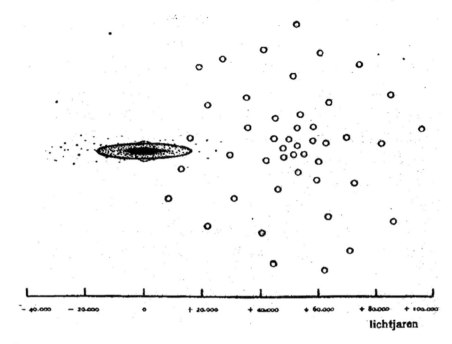

Fig. 12.7 The Kapteyn Universe and the system of globular clusters according to Shapley. This figure is attributed to Jan Hendrik Oort, and just like Fig. 6.6 is taken from the book *Kosmos* by Willem de Sitter, published in 1934

1913; Shapley had used this also to determine the distances of various globular clusters and found that they filled a roughly spherical volume.

The situation around 1922 was as shown in Fig. 12.7. The Kapteyn Universe is far too small in its long dimension, because the star counts from which it was derived are strongly affected by the absorption of starlight. But Shapley's system of globular clusters is far too large. He puts the center at 15 kpc and that is almost twice as far away as we now think. This is also an effect of absorption. However, the globular cluster M13 is in the sky far from the Milky Way and therefore out of the plane of the Kapteyn Universe, and there is little dust along the line towards it. But the center of Shapley's system lies in the Milky Way, and that is also influenced by absorption; the stars in the globular clusters appear fainter than they would have been without dust, so he estimated their distance far too great.

The situation in Fig. 12.7 is sometimes presented as a battle between Kapteyn and Shapley, especially in the United States, where Shapley would then be the ultimate winner: Kapteyn was wrong, because he neglected absorption by dust, and Shapley was right. But that is an oversimplification. Kapteyn worried about absorption, but had been persuaded that it could be neglected based on

Shapley's observations of no reddening towards M13; and the dimensions of Shapley's system are also wrong because of his assumption of no absorption towards globular clusters that on the sky appear near the Milky Way is also incorrect. Kapteyn may have had the dimensions in the Milky Way of his Universe completely wrong, but in two other respects he was surprisingly correct, namely the distribution of the luminosity of the stars near the Sun (his 'brightness curve') and the structure and dynamics in the direction perpendicular to the Milky Way.

12.5 Retirement and Mount Wilson

In 1920 the Kapteyns had to leave their upper floor at the Ossenmarkt, because the owner wanted to use it himself. Shortly after the war, there was also a serious shortage of housing in neutral Netherlands, but they could make use of two rooms in hotel De Doelen on the Grote Markt for a longer period of time. They were also allowed to take their own furniture there. In Hertzsprung–Kapteyn's biography [1] their rooms in the hotel were described as follows:

> The large living room at the front of the house, where the piano and the old-fashioned couch stood, overlooked the large market square with the stately Martini tower on the right and the massive City Hall on the left. On Tuesdays and Fridays it was Market Day, which brought with it a lot of bustle. One could sit for hours in the deep window-sill looking at to the bustling movement without getting tired. But madness broke out in the middle of the beautiful month of May, when then there was noise and shouting at their doorstep the fair with its merry-go-rounds and hippodromes, steam engines whistled and music thumped, and the balmy spring air was filled with the smell of pancakes and hot machines. And Groningen was partying, and it did not want to have their privileges taken away from them. That was a busy, unpleasant time for the guests of the hotel. The Kapteyns often took refuge and went elsewhere for peace and quiet, which they could not find at home.

Kapteyn kept up a busy schedule at the university (see Fig. 12.8). But at the end of the academic year in which Kapteyn turned seventy (1920–1921), to be precise on July 7, 1921, Kapteyn had to retire. He had agreed to a position as deputy director of the Sterrewacht Leiden for one day a week, but his work in Groningen was coming to an end. On his retirement Pieter van Rhijn (Fig. 12.9) was appointed as his successor as professor and director of the Astronomical Laboratory 'Kapteyn'. Kapteyn entrusted his students, particularly Jan Hendrik Oort and Egbert Adriaan Kreiken, in the hands of

Fig. 12.8 Photograph taken from Hertzsprung–Kapteyn's biography [1], en titled 'The professor and his secretary'

his successor. He was no longer actively involved in teaching and research and left Groningen immediately after his retirement. I let Henriette Hertzsprung–Kapteyn speak again [1]:

In June he gave his last lecture, and the Kapteyns said goodbye to Groningen. The good friends were invited in turns to their table in De Doelen. They found this a more intimate way of saying goodbye than during a joint big dinner, which offers more 'éclat' [lustre] than confidentiality. And when they said goodbye to the hotel, it was with melancholy for everything they were about to leave. The headmistress said goodbye in tears, saying that they had been the most pleasant and easy guests she had ever had. The appreciation was mutual, it had been good years. They left at seven o'clock in the morning. None of their friends knew about this departure, so no one escorted them out. That is how they wanted it, and in silence they said goodbye to the city, where they had lived and worked for 43 years. A long happy life of fruitful labor, of domestic happiness and of

Fig. 12.9 Pieter Johannes van Rhijn. Reproduction of a crayon drawing donated by his relatives to the Kapteyn Astronomical Institute. This drawing hangs in the Kapteyn Room

great friendships lay behind them, but a new and promising life lay ahead of this couple, that still had strength and zest to create new values in the final parts of their lives.

After the end of the First World War, Kapteyn had made plans to visit Mount Wilson again. He was still a Research Associate of the Carnegie Institution and still felt he had to deliver something of value to justify the money he received for it. In August 1919 he wrote George Hale. He asked Hale to excuse him for the year 1920, because of the much work he wanted to finish in Groningen before his retirement, but suggested that he would visit Pasadena and Mount Wilson again in 1921. The final catalog of the *Selected Areas*, based on the photographic work done with the 60 in. telescope, had not yet been completed. He wrote George Hale again in April 1921 with the proposition to arrive in California around December 1. He had a lot of work to do there, especially with Frederick Seares, but also with other staff members, with whom Kapteyn had become friends. One of them, whom we have not yet met, is Charles Edward St. John (1857–1935), who had joined the staff of Mount Wilson in 1908.

George Hale had problems with his stomach. In a letter also dated April 1921, which must have crossed Kapteyn's letter, Hale wrote that after much thought and after consultation with Seares and St. John, he had come to the conclusion that it was not wise for Kapteyn to come to Mount Wilson. Hale's health was such that he would not be working at the Observatory for some time, so he had to temporarily handed over the directorship to Walter Adams.

> I think that neither you nor Adams fully understood each other in the past, and I am very much afraid that for various reasons (such as differences of opinion about the 'Entente policy' [the Allies' attitude towards Germany]) misunderstandings may arise that could prevent or seriously disrupt your future cooperation with the Observatory.

Kapteyn finally gave up his intended visit. He wrote to Hale:

> I still think, as Adams apparently thinks, that without you in between us, misunderstandings can easily arise again. Which Heaven forbids. My attitude toward international cooperation (as you say) can only make it more difficult. So of course I will not come. And if you want to continue the cooperation with me – for which I am very grateful – I will do what I can from here.

12.6   The International Astronomical Union

In 1919 the International Astronomical Union IAU was founded. This was the ultimate body to facilitate international cooperation, of which Kapteyn had been a champion. However, in accordance with the practice of those days, Germany was excluded from participation and membership. Kapteyn therefore disapproved of the IAU.

The first General Assembly of the IAU took place in Rome in May 1922. At that time Kapteyn was too ill to attend and it is very likely that he had refused to participate anyway. The Committee for the IAU, which coordinated matters concerning the Union for the Netherlands, consisted of de Sitter, Hertzsprung, Nijland and Jacob Evert Baron de Vos van Steenwijk (1889–1978), who was secretary. The latter had a PhD in astronomy. He came from a family of politicians; after having been a teacher for some time, he became Mayor of Zwolle and later Haarlem, and Commissioner of the Queen in North Holland. However, he remained actively interested in astronomy. In Rome, a proposal was discussed by the Dutch delegation to take parallax plates of all 6188 stars in the *Preliminary General Catalog* every ten years. That sounds like something coming from the Kapteyn. However, he sent Willem de Sitter a postcard before the meeting: 'Amice, remember that the proposal you will defend is not a proposal from me to the Union. I want to have nothing to do with the Union. My view is that it is a proposal from the Dutch club. If you are looking for a name, it is Pieter van Rhijn, from whom the proposal is indeed coming.' A somewhat less far reaching proposal does indeed appear in the Congress report as having been adopted.

Kapteyn is not mentioned at all in this report, but apart from the above, there are three cases where his name would have been appropriate. The first concerns the reports on the progress of the *Plan of Selected Areas*; that Kapteyn's name was associated with it must have been obvious. The second is the resolution, adopted at the suggestion of the American 'Section' of the IAU (which worked under the auspices of the National Research Council), to define the absolute magnitude of stars as the apparent brightness at a distance of 10 parsec. This is the definition that Kapteyn had been using for years (and was the first to do so). Its widespread use was mentioned, but the name Kapteyn does not appear in the resolution or explanation.

Fig. 12.10 Edouard-Benjamin Baillaud, first President of the IAU. This picture comes from the *Album Amicorum*, presented to H.G. van de Sande Bakhuyzen on the occasion of his retirement as professor of astronomy and director of the Sterrewacht Leiden in 1908 [47]

Third, there was the opening speech of the first IAU president, Edouard-Benjamin Baillaud (1848–1934) (see Fig. 12.10). In his *in memoriam* on Kapteyn, Willem de Sitter wrote::

This modest workshop [the Astronomical Laboratory in Groningen] was the starting point for the research, which completely changed the appearance of astronomical science. Recently, Arthur Eddington, on the occasion of the Royal Astronomical Society's centenary celebration, listed the six most important astronomical events of these hundred years, including: 1904, discovery of two Star Streams by Kapteyn. And Baillaud, the veteran among astronomers, mentioned in his opening speech as president of the Congress in Rome in May of this year, the three things that revolutionized the face of science in the more than fifty years that he had worked as an active astronomer, and these three were: photography, the giant telescopes, and the Groningen Laboratory.

The latter is quite an exaggeration. Baillaud himself said (in my translation from French) [113]:

A new revolution is underway: after the invention of reflective telescopes, spectroscopy and photography, we are now witnessing the arrival of gigantic instruments [...]. These instruments often remain unproductive if they are not accompanied by all the necessary secondary devices found in an astronomical laboratory. Since they are photographic plates that need to be studied, these laboratories do not have to be next to the telescopes. The laboratory in Groningen provides excessive evidence of this.

12.7 The End

Kapteyn and his wife had intended to settle in Hilversum after his retirement. In preparation for this and to inform themselves as to houses there, they first temporarily moved in with their eldest daughter and son-in-law Noordenbos; the latter was now a professor in Amsterdam. Indeed they found a suitable home in Hilversum, but Kapteyn never lived there. His widow did live in that house for many years and daughter Henriette moved to the same street after she separated from Ejnar Hertzsprung in 1923.

Statistically speaking (for more details see my more extensive biography [5]) Kapteyn, at seventy had a further life expectancy of seven years. It was also true from relevant studies that academics who did not work in industry lived on average ten years longer than the general population, and astronomers and archaeologists another one and a half to two years [114, 115]. So he easily might have lived another decade. Unfortunately, it was not to be so. In the fall of 1921 the first signs appeared of the fatal disease that would kill him.

In January 1922 Kapteyn wrote to George Hale:

> My health has not been too good lately and I am writing in bed. After we left Groningen my wife and I were planning to travel for most of the year. We could do that without too much expense and it would give me time to rest after a very busy last year. We have been to Switzerland, various parts of Germany, Scotland and England and had a good time there. Since about the beginning of October, however, I did not feel well and after about a month in Leiden, where I was appointed adjunct-director (a position that takes about 1/7 of my time) we stayed in Amsterdam with our son-in-law, Prof. Noordenbos, where I entrusted myself to the hands of the medical profession. Fortunately nothing serious was found and the expectation is that soon I will feel normal again. Only the pains can go on for another month or more.

But it did not get any better. We do not know what Kapteyn was suffering from, but a form of cancer like Kahler's disease seems most likely. It occurs in the bone marrow and causes pain. Cancer was responsible for 15-20% of all deaths at the time.

Apparently Kapteyn himself was not aware of the seriousness of the situation, because in letters he continued to allude to a complete recovery. He no longer left the house of the Noorderbos family, where he was nursed until the end by his wife, his eldest daughter and his second daughter, who had been in Amsterdam permanently for the last month. After her husband's death, Mrs. Kapteyn told George Hale in a letter that her daughter (I presume the eldest, who was a physician herself) told her at the end of April that there was no chance of recovery (Fig. 12.11).

Kapteyn died in Amsterdam on 18 June 1922.

Henriette Hertzsprung–Kapteyn wrote in her biography [1]:

> Many true friends followed him to the Westerveld cemetery; no official homage, which the family had declined, expecting to act in his spirit, only true love and friendship accompanied him. Van Anrooy played on the organ the poignant closing choir from Bach's Saint Matthew Passion:
>
> > 'Wir setzen uns mit Tränen nieder
> > Und rufen dir im Grabe zu:
> > Ruhe sanfte, sanfte Ruh.'
>
> And his friend Bordewijk spoke loving words with an emotional voice. That was all, but the beautiful music for the friend, the deeply felt words, the silent sorrow

Fig. 12.11 Tombstones of the Kapteyn family at the 'Westerveld' cemetery at Driehuis in the municipality of Velsen. The middle one is for Jacobus Cornelius Kapteyn, his granddaughter Greta Noordenbos, his wife Catharina Elisabeth Kapteyn–Kalshoven and his son-in-law Willem Cornelis Noordenbos. On the left Kapteyn's daughter Jacoba Cornelia Noordenbos-Kapteyn, her son Willem Noordenbos and his wife. To the right Kapteyn's son Gerrit Jacobus and his wife Wilhelmina Henriette van Gorkom Provided by great-grandson Jan Willem Noordenbos, who looks after these graves.

of all fulfilled that hour with a holy consecration.

In the evening Kor Kuiler, the conductor of the Harmonie Orchestra, performed Beethoven's Funeral March as a tribute to the great Groninger.

Peter Gijsbert van Anrooy had been conductor of the Groninger Orkest Vereeniging, the local symphony orchestra of Groningen, (1905–1910) and at that period befriended Kapteyn. Hugo Willem Constantijn Bordewijk (1879–1939) was professor of law, economics and statistics between 1918 and 1938. He and his wife and the Kapteyns had become good friends in the years that the Kapteyns lived in Hotel de Doelen, not far from the Bordewijk residence.

Historian Johan Huizinga (1872–1945), son of Kapteyn's great friend Dirk Huizinga, had the intention to write a biography of Kapteyn together with Willem de Sitter (for more details see my scientific biography [5]), but that was never completed. In an in memoriam in the well-known literary and cultural magazine *De Gids* he wrote [116]:

A few weeks before his death, suffering but full of courage and hope, he told Peter van Anrooy, who in his time in Groningen was fortunate enough to become one of Kapteyn's best friends: Perhaps this is still the happiest time of my life. An extraordinarily fine observer, having come from abroad a short time earlier, once said to Willem de Sitter: 'What kind of a person was this Kapteyn? When you pronounce his name, something special comes into your voice and your face'. These are very simple things that I can tell about him, but one cannot speak of them other than very simply and unselfishly. When someone in his seventies, – was he really that old, that boyish, youthful one? – dies, then one usually testifies that he was able to give what had. In Kapteyn there were still treasures of labor and interest and love of mankind and things left. I had always imagined him working until his nineties. At Westerveld, last Tuesday, under Van Anrooy's organ play, I had to think of the spaces that this spirit has passed through, and how they have become for him now less than the size of his hand.

And finally I quote Eduard Jan Dijksterhuis (1892–1965). Dijksterhuis is known as a historian, who wrote the well-known book *The mechanization of the world picture* in 1950. He studied mathematics in Groningen (and was a teacher at an HBS for many years) and there he attended lectures by Kapteyn. He wrote in a magazine for secondary education [117]:

The whole scientific world honored Kapteyn for the great work he had done in the service of astronomy; his students also thanked him for the imperishable gift he had personally given them when he gave them the best that a teacher could give them, the love, admiration, and enthusiasm for the magnificent science to which he had dedicated his rich life.

Fig. 12.12 The widow Catherina Elisabeth (Elise) Kapteyn–Kalshoven in the 1930s. The book she holds in her hands is a copy of *J.C. Kapteyn; Zijn leven en werken* by Henriette Hertzsprung–Kapteyn [1]. Painting by Lizzy Ansingh (1875–1959), member of the group of painters 'the Amsterdamse Joffers'. Gift to the Kapteyn Astronomical Institute by great-granddaughter Wilhelmina Henriette de Zwaan-Kaars Sypesteyn for display in the Kapteyn Room

A

Some More Background

Someone told me that
each equation I included in the book
would halve the sales.
Stephen Hawking

As long as algebra is taught in school,
there will be prayer in school.
Cokie Roberts (née Mary Martha Corinne Morrison Claiborne Boggs)
(1943–2019)

This appendix aims at providing some additional background and understanding, but reading is not necessary for the reader not interested in such details. It avoids mathematical equations. An extended version with mathematical discussions can be found at my Kapteyn Website, for a direct download use www.astro.rug.nl/JCKapteyn/AppendixA.pdf.

A.1 Vibrating Flat Membranes

To give an idea of what Kapteyn's study of vibrating flat membranes in his PhD thesis entailed, I give here a short summary.

The questions addressed are related to the manner in which a membrane can vibrate, depending on shape and support. In those days it was not really

---

From: *A Brief History of Time* [118].
Cokie Roberts was an American journalist and author.

necessary to add much original work to the thesis and Kapteyn's is for a large part indeed a survey of the literature. But he does use observations that had been obtained by others.

Kapteyn wrote down the mathematical equations that govern the way membranes vibrate, which were first derived by Poisson. He then solved these for a few specific cases, such as a square, rectangular or circular membrane. For this he introduced boundary conditions that specify that the membrane cannot move at the edges and resulted in a predicted pattern of nodes, which are parts of the membrane that do not move. Like in the case of the string of a violin there is a fundamental tone which has only nodes at the two ends, a first overtone, in which the center of the string is also steady, a second overtone where there are nodes a third and two thirds along the length of the string, etc.

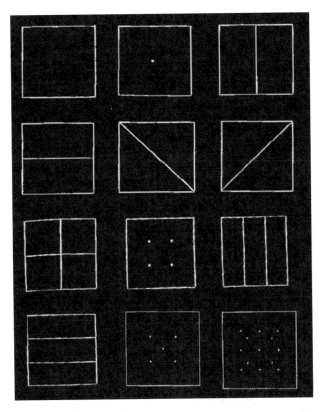

Fig. A.1 Linear nodes and point-like nodes for a vibrating square membrane according to the PhD thesis of Kapteyn. Top-left: wavelength twice the lengths of sides; then there is no node. Next we see some examples of 'overtones', where the wavelengths in each direction are equal to the length of the sides, two-thirds of this or half

Figure A.1 shows some of Kapteyn's results in terms of nodes. He finds that these are either point-like or straight lines. Kapteyn concludes that the point-like nodes must exist (apparently a new result), but cannot be seen in observations. This is probably due to small imperfections in the thickness and elasticity that in reality membranes have.

A.2   Distances and Luminosities

The distance of a star not too far from the Sun can be measured directly using the so-called annual parallax. In Fig. A.2 we see how the annual motion of the Earth around the Sun is reflected in an elliptical orbit of the star on the sky. The semi-major axis of that ellipse (which is equal to the angle p at the top of the triangle) then is a measure for the distance of the star. The radius of the Earth's orbit (the Astronomical Unit) is 1.4960×10^{11} m. When that angle, the parallax, is 1 arcsecond, the distance of the star is 3.0857×10^{16} m. This is called one *parsec*. It is equal to 3.26 lightyears (one lightyear is the distance traveled by light in vacuum during one year).

The apparent magnitude of a star is a measure of its brightness in the sky; the concept originates from Antiquity, when the brightest stars were assigned magnitude zero and the weakest that the human eye could see, magnitude six. This system was already used in the star catalog of Hipparchus of Nicaea in the second century BCE. The British astronomer Norman Robert Pogson redefined it in 1856 by proposing a scale where 5 magnitudes were exactly a factor of 100, so that one magnitude corresponds to a factor of $\sqrt[5]{100} = 2.512$.

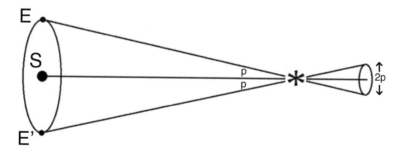

Fig. A.2   The annual circular motion of the Earth E around the Sun S on the sky in a projected elliptical motion by a star, of which the size of the ellipse ($2p$) depends on the distance to the star. The angles p are the parallax. Figure by the author

The absolute magnitude (designated with capital M) is defined as the magnitude a star would have at a distance op 10 parsec. This definition goes back to Kapteyn.

A.3   Sines Functions of Higher Orders

It is well known that the sine and cosine of an angle (in radians) can be expanded as an infinite series. The brothers Jacobus and Willem Kapteyn made a study of a set of generalized series. These are not new functions of angles but sets of mathematical formulae of which the usual sine and cosine are special cases. These series are therefore designated by 'higher-order-sines'. They published an extensive paper on this in 1886, that had been preceded by another one on a special case different from the sine and cosine.

A.4   Kepler's Equation

Planets, asteroids and comets move around the Sun in elliptical orbits according to the laws of Kepler. The geometry is illustrated in Fig. A.3. Suppose you want to find the position in the orbit at a certain time, then you need to know the angle v, which for historical reasons is called the 'true anomaly'. The object moves faster in its orbit when it is closer to the Sun, so v does not change uniformly with time. Kepler suggested the method to solve for v as shown in

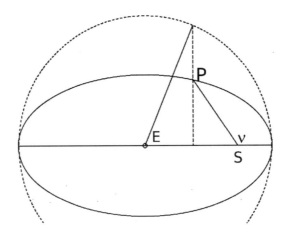

Fig. A.3 The orbit of a planet, asteroid or comet (P) is an ellipse with the Sun (S) in one of the foci. The true anomaly is v and the construction shows the definition of the eccentric anomaly E. For more explanation see the text. Figure by the author

Fig. A.3, where the angle E has been defined. This angle is called 'eccentric anomaly'. If you know E, v is easy to find from the equations for the ellipse. It does depend of course on the eccentricity of the orbit (0 for a circle, 1 for a line). But E also does not increase uniformly with time.

Therefore, Kepler defined an imaginary 'mean anomaly', M, which does run uniformly with time, but cannot be constructed in the figure. If you know the time t, M is easy to calculate. But then you have to find E and for that you have to use Kepler's equation, which connects M and E. This, however, is not straightforward to solve and had been a problem for studies of eccentric orbits of asteroids and particularly comets. This is a mathematical problem and ways had been found tot solve this, but it remained time consuming and tedious.

There were two general ways of tackling this problem. The first is by *iteration*, i.e. repeatedly calculating a better approximation until the result is sufficiently accurate. In that case you start with an initial guess of what M is for a particular value of E and insert it into the equation in such a way that an improved guess of M results. Insert that new value also until a sufficiently accurate value is obtained. If e is significantly different from 0, this may take a long time.

The second method is by *series expansion*, in which M is written as an infinite sum of terms with increasing powers of E, that decrease in value for increasing E. One then only needs to calculate a limited number of terms. Even for an asteroid in a fairly eccentric orbit, that may take a fair number of terms for the series expansions used.

The solution proposed by Kapteyn falls into the second category and consisted of a new series. However, the actual improvement was marginal.

A.5   Stellar Evolution

The fundamental diagram in astrophysics is named after the Danish astronomer Ejnar Hertzsprung and American Henri Norris Russell (see Fig. A.4). It is a diagram that relates the temperature at the surface of stars to the luminosity. The vertical axis is the luminosity, but for that also the absolute magnitude can be used. On the horizontal axis we have the temperature at the surface of the star, but one can also use for this the color index, for example the difference between the magnitudes of the star in a blue (B) and a visual (V) band. If the star is relatively bright in blue, then the star is relatively hot on the surface. But the spectral type can also be used for this, because the absorption lines in the spectrum of the star are created by atoms or ions in the outer parts, which absorb light at specific wavelengths; which atoms or ions are present and which lines are prominent, is strongly influenced by the temperature. Spectral types

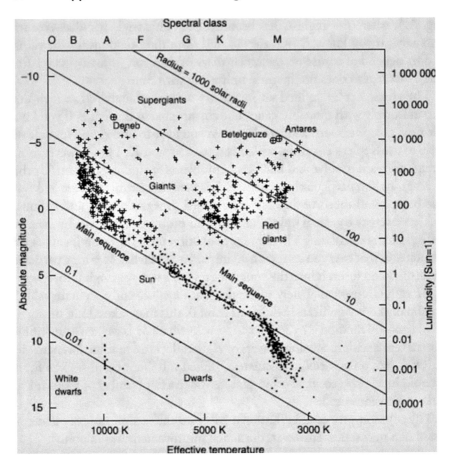

Fig. A.4 Hertzsprung–Russell diagram of stars. On the horizontal axis at the top spectral type and surface temperature in Kelvin. The vertical axis has on the left the absolute magnitude and on the right the luminosity expressed in that of the Sun. The lines are of constant radius expressed in solar radii. From [119]

are indicated by the letters $OBAFGKM$ and a decimal subdivision within that. By definition, the color index $(B - V)$ is zero for an $A0$-star. O-stars are relatively blue; along the sequence from top-left to bottom-right, the color becomes increasingly redder, while the temperature at the surface of the star decreases. In O-stars, the lines of ionized helium are strongest, in B-stars those of neutral helium, in A-stars those of hydrogen, while in F, G, and K-stars lines of ionized calcium and metals gradually become stronger. In M-stars, lines of molecules (e.g., titanium oxide) are strong.

The radiation of a star is so-called black-body radiation; a body of any temperature emits an amount of radiation integrated over all wavelengths, which depends only on the temperature T. The amount per unit surface area is given by the law of Stefan-Boltzmann, named after Josef Stefan (1835–1893) and Ludwig Eduard Boltzmann (1844–1906). When for a star its distance is known, we can calculate the total energy emitted and from the distribution of brightness with wavelength determine the temperature. This law then can be used to calculate the total surface area of the star and thus its radius.

Most stars lie along the Main Sequence from the top left to the bottom right. After being formed from a cloud of gas cloud, contraction and release of potential energy will make the inside of a star hotter, and as the density increases more energy is added than can be radiated away. At ten million degrees, nuclear reactions start, converting hydrogen into helium. The stars at the top are bright, heavy, and hot and for them this period is shortest, the M-stars are faint, light and cool and they live longer than the present age of the Universe.

Eventually all hydrogen in the central parts of a star will be used up, so that the core, which now consists entirely of helium, will be extinguished. Initially 'hydrogen burning' continues for a while in a shell around the core and the star becomes brighter. A star like the Sun then moves up along the sub- and the giant branches (along numbers IV and III in Fig. A.4). The temperature T at the surface drops and the star becomes redder, but the radius R will then increase. Because energy is no longer produced there, the core contracts, but then gets hotter. When it becomes hot enough, helium burning will start there, converting it into carbon and oxygen. A star like the Sun is then on the 'clump' halfway along the red giant branch. Then the star expels its outer layers and forms a so-called planetary nebula. The star then cools down to a white dwarf at the bottom of the figure. The matter becomes very compact, but at a certain moment the contraction stops because Fermi's exclusion principle and Heisenberg's uncertainty relationship together forbid that the electrons come even closer together (the so-called degeneration pressure). The uncertainty relationship says that the position and speed of an electron (or other elementary particle) cannot be measured infinitely accurately at the same time. Then two different electrons can in principle not get so close together and have the same speed so accurately that they would be indistinguishable. The Fermi principle prohibits this and so matter cannot become more compact than when all particles can just be distinguished in position and speed. This is called degeneration pressure.

In a more massive star than the Sun, the pressure in the core becomes so great that the center continues to contract; the star then becomes so hot that even heavier chemical elements are formed. But in the long run this process also stops and the central parts contract even further. The contraction force then is so great that the degeneration pressure can no longer compensate for it; then the electrons are, as it were, merged into the protons and form neutrons. This process is so fast that an enormous pressure wave propagates through the star, blowing itself up like a supernova. More chemical elements are formed and this material is thrown out into space. The remaining central parts are then a neutron star, which is in equilibrium under the influence of the degeneration pressure of the neutrons, or, if the initial star mass is even larger and the contraction forces too strong, a black hole.

The derivation of the formulae above and more background on the issues in this section can be found among others in my introductory astronomy lectures [119, 120].

A.6   The 'Star Ratio'

This concept was used by Edward Pickering in 1903 to obtain information about the distribution of stars in space. Assume that all stars have the same intrinsic luminosity and are uniformly distributed in space. Take a certain distance from the Sun, then all stars at that distance have the same apparent brightness or magnitude, say m. Stars of a magnitude weaker, i.e. $m + 1$, in the sky are according to the definition of the magnitude scale a factor $\sqrt[5]{100} = 2.512$ fainter and have to be therefore a factor $\sqrt{2.512}$ further away. The stars between apparent magnitude m and $m + 1$ fill a shell. Now take stars that are another magnitude fainter, so $m + 2$. They are a factor of 2.512 fainter than the stars of magnitude $m + 1$ and are a factor of $\sqrt{2.512}$ further away. The shell between $m + 1$ and $m + 2$ is a factor 2.512 larger than the one between m and $m + 1$ and the thickness is a factor $\sqrt{2.512}$ larger. So the volume of the shell is a factor $2.512 \times \sqrt{2.512} = 2.512^{3/2}$ larger and so is the number of stars in it.

This means that the number of stars between magnitude $m + 1$ and $m + 2$ should be the same factor $2.512^{3/2} = 3.981$ times larger than the number between m and $m + 1$. This ratio was called the 'star ratio' by Edward Pickering and 3.981 is its reference value. In practice, he did not use the factor 3.981 itself, but actually its logarithm, which is exactly 0.6. In reality on the sky, this ratio is smaller than this theoretical value; this may be due to changes in the density of the stars with changing distances or to the distribution of their intrinsic luminosity, but also to absorption of starlight by dust in interstellar space.

A.7   Skew Probability Distributions

An example of a skew probability distribution that Kapteyn studied, is the so-called 'log-normal' distribution. This is similar to the well-known Gaussian probability distribution, but now when plotted on a logarithmic horizontal scale. An example of a log-normal distribution has been drawn on the left in Fig. A.5. To the right the same distribution on a logarithmic scale on the x axis, so that it becomes the normal Gaussian distribution. The color changes with integer values of the 'dispersion' σ.

In normal distribution, the mode (the maximum or most common value), the median (half larger and half smaller) and the regular average have the same value (in the figure 10^2 or 100). At the log-normal, the average is 127.8, but the mode is at 62.2 and the median at 100.

Kapteyn went on to study a general family of skewed distributions, of which this is only one example. An extensive, technical discussion of Kapteyn's work in the field of statistics has been published an article by Ida H. Stamhuis and Eugene Seneta [96], but see also [97].

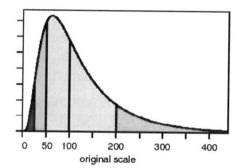
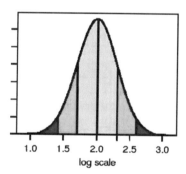

Fig. A.5 On the left a log-normal probability distributions. On the right the same distribution but plotted on a logarithmic horizontal scale. After [97]

B

Literature

One should distrust the compulsion to write,
but not the urge to formulate.
Godfried Bomans

The secret to creativity is
knowing how to hide your sources.
Albert Einstein

There are many publications about Kapteyn that can serve as suggestions for further reading, partly in Dutch, but for the most part in English. More details and in many cases links to electronic versions of these can be found on my special Kapteyn Webpage www.astro.rug.nl/JCKapteyn. I will present an overview here.

• First of all, there are five publications that I already mentioned in the Preface. Of these *Persoonlijke herinneringen aan J.C. Kapteyn* (Personal memories of J.C. Kapteyn) by Cornelis Easton, *J.C. Kapteyn: Zijn leven en werken* (J.C. Kapteyn: His life and work) by Henriette Hertzsprung–Kapteyn and *Lieve Lize: De minnebrieven van de Groningse astronoom J.C. Kapteyn aan Elise Kalshoven, 1878–1879* (Dear Lize: The love letters of the Groningen astronomer J.C. Kapteyn to Elise Kalshoven) by Klaas van Berkel and Annelies Noordhof-Hoorn are very accessible to a wide Dutch audience. For English speakers the symposium proceedings *The Legacy of J.C. Kapteyn; Kapteyn and the development of modern astronomy* by myself and Klaas van Berkel and my extensive biography *Jacobus*

P. C. van der Kruit, *Pioneer of Galactic Astronomy: A Biography of Jacobus C. Kapteyn*, Springer Biographies, https://doi.org/10.1007/978-3-030-55423-1

Cornelius Kapteyn; Born investigator of the heavens are recommended; these contain a lot of details and deal with technical matters regarding his scientific research.

• Various articles have been written about Kapteyn over the years, sometimes as an *in memoriam* shortly after his death or as a eulogy at the presentation of honors or as short biographies. These can be found on the Kapteyn Webpage mentioned above, where generally links are available to electronic copies. Kapteyn's scientific genealogy is also presented there.

• Most of Kapteyn's letters and other archival material have been lost; probably during the bombardment of Rotterdam in May 1940. This can probably be explained as follows (see also the discussions in the last two of the works just mentioned): Willem de Sitter and Johan Huizinga intended to write a biography of Kapteyn, but never realized it. Aernout de Sitter, son of Willem, might have had the same intention later; when he became director of the Observatory of Lembang in the former Dutch East Indies, he could then have shipped the Kapteyn archives in a large chest, which happened to be located in the Rotterdam harbor in 1940. Fortunately, but without any real apparent reason, the letters from David Gill to Kapteyn – and only those letters – have been preserved in the archives of the Kapteyn Astronomical Institute, University of Groningen. I have scanned them and made them publicly available in electronic form on the Kapteyn Webpage.

• Kapteyn's scientific publications are for a large part available electronically on the Astronomy Data System (ADS; see below). References and links can be found on the Kapteyn Webpage; where articles are not available in ADS, I have in some cases (especially Dutch language popular articles) made them available there electronically.

Astronomy is in the fortunate circumstance that almost all of the professional literature, as far as it has appeared in scientific (and also many popular) journals or as observatory or institute publications or (annual) reports, has been scanned and made available electronically. If applicable, this even goes back to the eighteenth or nineteenth century. This project, the Harvard–Smithsonian Center for Astrophysics/National Aeronautics and Space Agency Astrophysics Data System (SAO/NASA ADS, or ADS for short) is available free of charge to anyone up to the year in which magazine publishers have introduced electronic subscriptions. In the latter case, a link to the magazine's site is provided; sometimes it is available in the framework of Open Access, but in some cases such a subscription is be required. Recently, for most astronomical journals, the articles are made available one year after publication as part of Open

Access and are therefore also available to everyone free of charge. The URL is iu.adsads.harvard.edu. Search functions, citation scores and bibliometric indicators are also available in ADS.

Articles are identified in ADS by a code that contains year, journal or otherwise, first page and first letter of the author's name. So Kapteyn's article 'First attempt at a theory of the arrangement and motions of the Sidereal System' in the *Astrophysical Journal*, volume 55, pages 302–328 (1922), has ADS code 1922ApJ....55..302K. It can then be found on the Web with ui.adsabs.harvard.edu/abs/1922ApJ....55..302K/abstract.

Kapteyn in 1908 in the 'Monastery' (the living quarters astronomers use during their observing sessions) at Mount Wilson Observatory. Courtesy the Observatories of the Carnegie Institution for Science Collection, Huntington Library, San Marino, California [121]

References

1. *J.C. Kapteyn; Zijn leven en werken*, by H. Hertzsprung-Kapteyn, Wolters 1928, www.dbnl.org/tekst/hert042jcka01_01/, www.astro.rug.nl/JCKapteyn? HHKbiog.html.

Preface

2. Cited in *Contraterrein* by R.A. Cornets de Groot. [www.cornetsdegtoot.com/pdf/Contraterrein.pdf]
3. *The life and works of J.C. Kapteyn: An annotated translation with preface and introduction*, by E. Robert Paul, Kluwer (1993), ISBN 978-07-923-2603-8.
4. *The legacy of J.C. Kapteyn: Studies on Kapteyn and the development of modern astronomy*, by P.C. van der Kruit & K. van Berkel, Springer, ISBN 0-7923-6393-0 (2000).
5. *Jacobus Cornelius Kapteyn: Born investigator of the Heavens*, by P.C. van der Kruit, Astrophysics and Space Science Library, Vol. 416, Springer Publishers, ISBN 978-3-319-10875-9 (2015). www.astro.rug.nl/JCKapteyn.
6. *De inrichting van de hemel: Een biografie van astronoom Jacobus C. Kapteyn*, by P.C. van der Kruit, Amsterdam University Press (2016), ISBN 978-94-6298-042-6.
7. *Jan Hendrik Oort: Master of the Galactic System*, by P.C. van der Kruit, Astrophysics and Space Science Library, Vol. 459, Springer Publishers, ISBN 978-3-319-10875-9 (2019). Accompanying Website at www.astro.rug.nl/JHOort.

© The Editor(s) (if applicable) and The Author(s), under exclusive license to Springer Nature Switzerland AG 2021
P. C. van der Kruit, *Pioneer of Galactic Astronomy: A Biography of Jacobus C. Kapteyn*, Springer Biographies, https://doi.org/10.1007/978-3-030-55423-1

8. *Horizonnen*, by P.C. van der Kruit, Prometheus Amsterdam, ISBN 798 90 446 4144 8 (2020). English translation *Master of Galactic astronomy: a biography of Jan Hendrik Oort*, to appear in Springer biographies, Springer Publishers ISBN 978-3-030-55547-4 (2021).

9. *Pesonal memories of J.C. Kapteyn*, by C. Easton, electronically available (in Dutch) through my Kapteyn Website www.astro.rug.nl/JCKapteyn/Easton.pdf, Appendix C. in [5], https://link.springer.com/content/pdf/bbm%3A978-3-319-10876-6%2F1.pdf.

10. *Lieve Lize: De minnebrieven van de Groningse astronoom J.C. Kapteyn aan Elise Kalshoven, 1878–1879*, by K. van Berkel and A. Noordhof-Hoorn. University of Groningen (2008), ISBN 978-90-367-3353-3.

11. *Ancestry*, www.ancestry.com/genealogy.

12. *Pondes*, www.pondes.nl.

13. *Geni*: www.geni.com.

14. *Online Familieberichten*, www.online-familieberichten.nl.

15. *Stamboomzoeker*, www.stamboomzoeker.nl.

16. *GenVer*, www.genver.nl.

17. *Delpher*, www.delpher.nl.

18. *Internationaal Instituut voor Sociale Geschiedenis*, www.iisg.nl/hpw/calculate.php.

19. *Currencyconverter*, www.historicalstatistics.org/Currencyconverter.html

20. *Dutch Ancestry Coach*, www.dutchancestrycoach.com/content/how-much-did-you-say-converting-dutch-historic-currencies (discontinued).

1. Prelude and Framework

21. www.clker.com/clipart-kaart-nederland-jan.html; public domain clip art.

22. www.taalwinkel.nl/playlist/presenteren-2/.

23. www.brainyquote.com/quotes/quotes/a/alnakay451982.html?src=t_context.

24. Nr.   03.0934;   https://beeldbank.ede.nl/atlantispubliek/default.aspx? modules=beeldbanken#media/434609/1.

25. *De ontdekkers van de hemel*, by D.M. Baneke, Prometheus, ISBN 978-90-351-3688-5 (2015).

26. www.genver.nl.; Gelderland, Barneveld, file 5808804 (Geboorten 1849-1852), image 108/210.

2. Boarding School and University

27. *Waarom daar?*, by J.D. Groen, in *Refogechiedenis in perspectief*, F. van Lieburg (red.), Uitgeverij Groen, ISBN 978-90-5829-780-8 (2007). www.dutchbiblebelt.org/fileadmin/Uploads/2007Refogeschiedenis_BUNDEL.pdf.

28. *Aart Jan Theodorus Jonker*, by M. van Rhijn, H.J. Paris Amsterdam (1929), p. 172.
29. https://ru.wikipedia.org/wiki/%D0%A4%D0%B0%D0%B9%D0%BB: Zoellner_Photometer.png.
30. *The trouble with the instrument: Zöllner's Photometer*, by K.B. Staubermann, Journal for the History of Astronomy, 31, 323 (2000), [2000JHA....31..323S].
31. *Karl Friedrich Zöllner and the historical dimension of astronomical photometry*, by C. Sterken & K.B. Staubermann, eds., VUB University Press, Brussels, SBN-13: 978-90-548-7254-2 (2000).

3. Astronomy Around 1875

32. Cited by Agnes Mary Clerke (1842–1907) in *Dictionary of National Biography*, Vol. 26. Available at https://en.wikisource.org/wiki/Herschel,_William_ (DNB00).
33. www.eso.org/public/images/potw1034a/.
34. https://commons.wikimedia.org/wiki/File:Johannes_Hevelius_und_ Elisabetha_Hevelius.jpg (copyright expired).
35. www.eso.org/public/images/yb_southern_cross_cc/.
36. *Time and time again: Determination of longitude at sea in the 17th Century*, by R. de Grijs, Institute of Physics Publishing, ISBN 9-780-750311-953 (2017).
37. *Early star catalogues of the southern sky. De Houtman, Kepler (second and third classes), and Halley* door F. Verbunt & R.H. van Gent, Astronomy and Astrophysics, 530, A93 (2011). [ADS: 2011A&A...530A..93V]
38. Archives Sterrewacht Leiden.
39. https://commons.wikimedia.org/wiki/File:PSM_V09_D079_Herschel_40_ foot_telescope_at_slough.jpg. (copyright expired).
40. *On the construction of the Heavens*, by F.W. Herschel, Philosophical Transactions of the Royal Society of London, 75, 213-266 (1785). [ADS: 1785RSPT...75..213H]

4. Observator and Professor

41. Jaarboek der Rijksuniversiteit te Groningen, 1878–1879, https://archive.org/ details/jaarboekderrijk00grongoog.
42. https://nl.m.wikipedia.org/wiki/Bestand:Prof._H.G._van_de_Sande_ Bakhuyzen.jpg.
43. www.rug.nl/museum/galerij/portretten/hoogleraar/mulerius.
44. *Sterrenkijken bekeken*, door A. Blaauw, J.A. de Boer, E. Dekker & J. Schuller tot Peursum-Meijer, University Museum Groningen (1983).
45. www.rug.nl/museum/galerij/portretten/hoogleraar/mees.

46. https://kaarten.abc.ub.rug.nl/root/grs/krt-1868-grs/ and https://atlas1868.nl/gr/groningen2.html.

5. Nearly Half a Million Stars

47. https://local.strw.leidenuniv.nl/album1908/book_info.html.
48. Museum Boerhaave, www.museumboerhaave.nl/object/universeelinstrument-v10680/.
49. Karl Ludwig von Littrow (1811–1877), in: *Sitzungsberichte der Mathematisch-Naturwissenschaftlichen Classe der Kaiserlichen Akademie der Wissenschaften*, Abtheilung II (Mathematik, Physik, Chemie, Mechanik, Meteorologie und Astronomie), LII, 546-546 (1866), https://catalog.hathitrust.org/Record/003086683.
50. www.saao.ac.za/about/history/.
51. *Einstein's Clocks, Poincaré's Maps, Empires of Time* by P. Galison, Norton, Londen; ISBN 0-393-02001-0 (2003).
52. *Digital Sky Survey* (DSS): https://archive.stsci.edu/cgi-bin/dss_form.
53. *Sloan Digital Sky Survey* (SDSS): www.sdds.org/dr12.

6. Laboratory and Statistical Astronomy

54. From: *Lijk Halfstok* by Havank, Bruna, ISBN 9044930702 (1948), p.1:
55. From: *Leven en werk van M.C. Escher*, by F.H. Bool, J.R. Kist, J.L. Locher & F. Wierda, Meulenhof, ISBN 9029011477 (1981), p.113.
56. *The Hertzsprung Russell Diagram by Richard Powell*, www.atlasoftheuniverse.com/hr.html and https://en.wikipedia.org/wiki/File:HRDiagram.png.
57. www.eso.org/public/images/eso0932a/.
58. ESA/Hubble & Digitized Sky Survey 2. Davide De Martin (ESA/Hubble), https://spacetelescope.org/images/heic1112f/.
59. National Optical Astronomy Observatories. Dale Cupp & Flynn Haase/NOAO/AURA/NSF/. www.noao.edu/outreach/aop/observers/n891.html.
60. www.astro.rug.nl/~vdkruit/Courses.html.
61. *Digital Sky Survey DSS2*, Space Telescope Science Institute: https://stdatu.stsci.edu/dss/.
62. *Willem de Sitter: Einstein's friend and opponent* by J. Guichelaar, Springer biographies, Springer Publishers, ISBN978-3-319-98336-3 (20018).
63. *Resumé van K's onderzoekingen naar de 'Structure of the Universe'*, 12 April 1907, www.astro.rug.nl/JCKapteyn/byJCK.html.
64. The International Astronomical Union Constellations: www.iau.org/public/themes/constellations/.

65. Koninklijk Natuurkundig Genootschap, www.kng-groningen.nl.
66. *Een Spiegel der Wetenschap: 200 jaar Koninklijk Natuurkundig Genootschap te Groningen*, Profiel, ISBN 90-529-4220-X (2001).
67. *Zekerheden in Waarnemingen: Natuurwetenschappelijke ontwikkelingen in Nederland rond 1900*, by J. Guichelaar , G.B. Huitema and H. de Jong, Verloren, ISBN 97-8908-704-194-6 (2012).
68. www.rug.nl/museum/galerij/portretten/hoogleraar/huizinga.

7. Star Streams

69. https://194.171.109.12/cat_toon_foto.php? registratiecode=VFOTNL029009&exact=JA&cat=VFOT& zoekterm=Kapteyn.
70. From *International Congress of Arts and Science*, Vol. VIII: Astronomy and Earth Sciences, red. Howard J. Rogers (1908), https://archive.org/details/internationalcon08inteiala.
71. Wikimedia Commons, https://commons.wikimedia.org/wiki/File:Karl_schwarzschild.jpg. Copyright expired.
72. www.noao.edu/image_gallery/html/im0552.html. Credit line: National Optical Astronomy Observatory/Association of Universities for Research in Astronomy/National Science Foundation.
73. www.spacetelescope.org/images/heic0515c/.
74. From: *Convergent of a moving cluster in Taurus* by Lewis Boss, Astronomical Journal, 26, 31-36 (1908), [ADS: 1908AJ.....26...31B].
75. https://bookhistory.harvard.edu/takenote/sites/default/files/attachments/EdwardPickering.jpg.

8. In the Meantime in Groningen

76. In: *Universitas Groningana MCMXIV–MCMLXIV*, Wolters, 135–188 (1914).
77. *Some notes on my life as an astronomer*, by J.H. Oort, Annual Review of Astronomy & Astrophysics, 19, 1-5 (1981). [ADS: 1981ARA&A..19....1O].

9. Mount Wilson

78. https://hdl.huntington.org/cdm/singleitem/collection/p15150coll2/id/399/rec/69.
79. https://hdl.huntington.org/cdm/singleitem/collection/p15150coll2/id/431/rec/91.

80. https://hdl.huntington.org/cdm/singleitem/collection/p15150coll2/id/711/rec/5.

81. T. Credner & S. Kohle, www.allthesky.com.

82. European Southern Observatory Press Release ESO1103, January 2011. www.eso.org/public/news/eso1103/.

83. News Release: STScI-2004-20. https://hubblesite.org/newscenter/archive/releases/2004/20/image/a/''/.

84. www.spacetelescope.org/projects/fits_liberator/fitsimages/john_corban_4/.

85. ESO Press Release ESO9934, February 1999. www.eso.org/public/news/eso9934/.

86. https://en.wikipedia.org/wiki/Messier_13#/media/File:M13_from_an_8_SCT.jpg.

87. https://hdl.huntington.org/cdm/singleitem/collection/p15150coll2/id/1124/rec/20.

88. *The Fourth Conference of the International Union for Co-operation in Solar Research*, Popular Astronomy, 18, 489-503 (1910). ADS: 1910PA.....18..489W.

89. https://hdl.huntington.org/cdm/singleitem/collection/p15150coll2/id/855/rec/3.

90. Susanne Elisabeth Nørskov, AU Library, Fysik & Steno, Institut for Fysik og Astronomi, Aarhus Universitet, Denmark.

10. Statistics and Other Concerns

91. https://nl.wikipedia.org/wiki/Ossenmarkt_%28Groningen%29#/media/File:Ossenmarkt_6_-_18613.jpg.

92. *Skew frequency curves in biology and statistics, 2nd paper*, by J.C. Kapteyn & M.J. van Uven, Hoitsema Brothers, Groningen (1916).

93. *Replicating the lost skew distribution machine of Jacobus Cornelius Kapteyn*, by F.L.R. Lucas & S.A.M. van der Salm, In: Calculating in everyday life, ed. K. Kleine, Deutsche Nationalbibliothek, ISBN 978-3-7448-1056-2, p.231 (2017).

94. https://hoogleraren.ub.rug.nl/?page=showPerson&type=hoogleraar&hoogleraar_id=108&lang=en.

95. https://hoogleraren.ub.rug.nl/?page=showPerson&type=hoogleraar&hoogleraar_id=125&lang=en.

96. *Pearson's statistics in the Netherlands and the Astronomer Kapteyn*, by I. Stamhuis and E. Seneta, International Statistical Review 77, 96-117 (2009).

97. *Lognormal distributions across the sciences: Keys and clues*, by E. Limpert, W.A. Stahel & M. Abbt, BioScience 51, 341-352 (2001).

98. Foto Groninger Archieven (1986-446); www.beeldbankgroningen.nl/beeldbank/, NL-GnGRA_1986_446.

99. *Herinneringen: Herinneringen uit de arbeidersbeweging: Sterrenkundige herinner-ingen,* by Anton Pannekoek, B.A. Sijes, E.P.J. van den Heuvel, J.M. Welcker & J.R. van der Leeuw, van Gennep, ISBN 978-90-60123-50-8 (1982).

11. The Kapteyn Universe

100. www.ocf.berkeley.edu/~fricke/dyson.html.
101. *Origins of the expanding Universe: 1912-1932,* by M.J. Way & D. Hunter (eds.), Astronomical Society of the Pacific Conference Proceedings, 471, (2013), [ADS: 2013ASPC..471.....W].
102. https://hdl.huntington.org/cdm/singleitem/collection/p15150coll2/id/1868/rec/26,
103. NASA, ESA en de Hubble Heritage Team (STScI/AURA), https://heritage.stsci.edu/2013/13/index.html.
104. *Egbert Adriaan Kreiken: pionier in de Turkse sterrenkunde,* by Juus Kreiken, Zenit, January 2013, 14-18.

12. Coda

105. *Who Discovered the Galaxy,* by H.A. Couper, British Astronomical Assocociation Journal, Vol. 96, No. 5 (1986), p. 293, [ADS: 1986JBAA...96..284C].
106. *The Interallied Conference of Scientific Academies in London (Translation),* by A.G. Marshall, Publications of the Astronomical Society of the Pacific, 30, 331-335 (1918). [ADS: 1918PASP...30..331M]
107. *To the members of the academies of the allied nations and of the United States of America,* by J.C. Kapteyn & G. Heymans, www.astro.rug.nl/JCKapteyn/Statement_USAcad.pdf.
108. *De stem van de wetenschap; Geschiedenis van de Koninklijke Nederlandse Academie van Wetenschappen,* by K. van Berkel Vol. 2, 1914-1918, Bert Bakker, ISBN 978-90-351-360104 (2011).
109. *Als bij toverslag; De reorganisatie van de Leidse 5terrewacht, 1918-1924,* by D.M. Baneke, BMGN/Low Countries Historical Review, 120, 207-225 (2005).
110. *Hij kan toch moeilijk de sterren in de war schoppen. De afwijzing van Pannekoek als adjunct-directeur van de Leidse Sterrewacht,* by D.M. Baneke, Gewina, 27, 1-13 (2004).
111. *Sterrekundig Instituut Utrecht: The last years,* by C.U. Keller, In: 370 Years of astronomy in Utrecht, ASP Conference Series, Vol. 470. p.3 (2013). [ASDS: 2013ASPC..470....3K] or arxiv.org/pdf/1208.4095.pdf.
112. www.pourlemerite.org.

113. *Transactions of the International Astronomical Union*, Vol. I: First General Assembly held at Rome, May 2nd to May 10th, 1922, ed. A. Fowler, London, Imperial College Bookstall.

114. *Average age at death of scientists in various specialties*, by S.M. Luria, Public Health Reports 84, 661–664 (1969), www.ncbi.nlm.nih.gov/pmc/articles/PMC2031509/.

115. *On the life expectancy of astronomers*, by D.B. Herrmann, The Messenger, 67, 62-63 (1992), [ADS: 1992Msngr..67...62H].

116. J. Huizinga in *De Gids*, jaargang 86. 130-133, 1922, www.dbnl.org/tekst/_gid001192201_01/_gid001192201_01_0081.php.

117. E.J. Dijksterhuis, in *Weekblad voor Gymnasiaal en Middelbaar Onderwijs*, 18. No. 42, 1909-1910, 1922.

A. Some More Background

118. *A Brief History of Time: From the Big Bang to Black Holes*, by S.W. Hawking, Bantam Dell, ISBN 978-0-553-10953-5 (1988).

119. *Fundamental astronomy*, by Hannu Karttunen, Pekka Kröger, Heikki Oja et al., Springer Nature, https://link.springer.com/book/10.1007/978-3-662-53045-0.

120. *Introductory astronomy*, by P.C. van der Kruit www.astro.rug.nl/~vdkruit/Inleiding.html.

B. Literature

121. https://hdl.huntington.org/cdm/singleitem/collection/p15150coll2/id/569/rec/3,

Index

© The Editor(s) (if applicable) and The Author(s), under exclusive license to Springer Nature Switzerland AG 2021
P. C. van der Kruit, *Pioneer of Galactic Astronomy: A Biography of Jacobus C. Kapteyn*, Springer Biographies, https://doi.org/10.1007/978-3-030-55423-1

297

Printed in the United States
by Baker & Taylor Publisher Services